CALIFORNIA
INSECTS

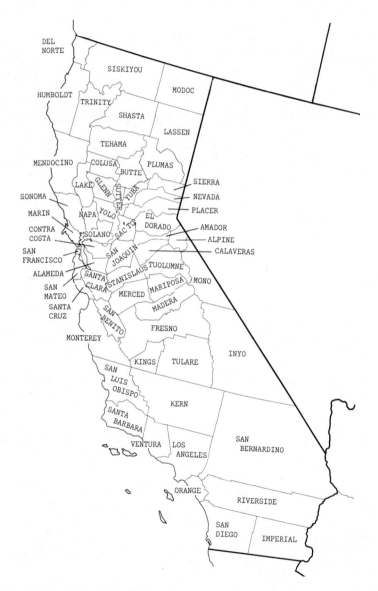

Map 1 Outline map of California, showing counties.

History Guides: 44

CALIFORNIA INSECTS

BY

JERRY A. POWELL

AND

CHARLES L. HOGUE

DRAWINGS BY
CHARLES L. HOGUE

UNIVERSITY OF CALIFORNIA PRESS
Berkeley   Los Angeles   London

California Natural History Guides
Arthur C. Smith, General Editor

Advisory Editorial Committee:
Raymond F. Dasmann
Mary Lee Jefferds
A. Starker Leopold
Robert Ornduff
Robert C. Stebbins

University of California Press
Berkeley and Los Angeles, California

University of California Press, Ltd.
London, England

CONTENTS

PREFACE

Few people appreciate the diversity of insect life, and most naturalists are surprised, even skeptical, when they hear estimated numbers of insect species in a given natural community or geographical region. There are about one million described species of insects in the world, more than 5 times the number of all other animals combined. Estimates of the number remaining to be discovered and named vary from 1.5 to 5 million or more. Discrepancies in such guesses are due primarily to the poor state of knowledge concerning tropical communities. Even in California, discovery of new species is commonplace; with each family inventoried by specialists during the past twenty years, 10–40 percent of the species have been discovered to be new. Only with quite well-known, large insects that are popular with collectors, such as dragonflies, butterflies, and large moths, are we relatively sure that the inventory is nearing completion. For example, 3 new species of butterflies have been described from California in the past two decades. There is no list of all insect species recorded in the state, but by comparing the numbers in several families recently treated in *The Bulletin of the California Insect Survey* with their counterparts in state lists of New York and Connecticut, a figure of 27–28,000 species emerges as a conservative estimate for California.

Therefore, it is obvious that a small book cannot treat all California insects. Whereas pocket guides can be developed with species treatments of California's birds (260 species), amphibians and reptiles (123 species), cacti (54 species), etc., comparable treatment of insects would fill many volumes, a prospect that is neither practical in terms of cost, nor possible in terms of available knowledge. In keeping with the philosophy that often prevails in outfitting expeditions or appointing museum staffs — 1 mammalogist, 1 ornithologist, 1 herpetologist, 1 ichthyologist, and 1 entomologist — one book on insect species of California will have to suffice.

Therefore a primary problem in planning this work was encountered at the outset: How does one summarize knowledge on 28,000 species of insects? We decided to select about 600 of the most common and interesting kinds and to discuss the biology, natural history, and diagnostic features of each, letting these represent the whole diversity. Smaller, more obscure groups are characterized at the family level, while more conspicuous insects and those of importance to man are treated in more detail with representative species or genera. Each of these entities is illustrated with one or more drawings or color photographs.

This book is written as a guide to identification and natural history of insects and is for the general public and beginning students of entomology, the study of insects. Selection of insects included was based on the kinds most often brought by the urban public to the Los Angeles County Museum of Natural History, on species most often collected by students in a field entomology course at the University of California, Berkeley, and on our combined field experience of more than fifty years of collecting in California.

Our identifications of species and statements on their natural history, biology, and geographical distributions are based on literature accounts (see "References" following the introduction to each order and "Learning More about Insects" at the end of the book); on collections of the California Academy of Sciences, San Francisco, Los Angeles County Museum of Natural History, and University of California, Berkeley; and on advice from our colleagues. No doubt readers will find conspicuous insects common in a given region, yet omitted from this treatment.

Intended as an introductory guide to identification and natural history of California insects, this book does not attempt to compile all information on the species discussed or to present technical information on other fields of entomology. Another book in the Natural History Guide series is scheduled which will discuss insect structure, physiology, growth and metamorphosis, ecology, classification, insects' relationships to man, and insect study. Therefore, these aspects are omitted or given only cursory mention in our treatment.

ACKNOWLEDGMENTS

We are indebted to specialists who reviewed individual chapters for errors of fact: Peter Bellinger (Collembola), Richard Allen (Ephemeroptera), Rosser Garrison (Odonata), Richard Baumann (Plecoptera), David Weissman (Orthoptera), Sigurd Szerlip (Heteroptera), Philip Adams (Neuroptera), John Doyen (Coleoptera), Vincent Resh (Trichoptera), Julian Donahue (Lepidoptera), Paul Arnaud (Diptera), Don Rohe (Siphonaptera), and Roy Snelling (Hymenoptera).

Many other specialists provided information or assisted by commenting on particular aspects of the text, including the following: George Buxton, John Chemsak, Michael Collins, Donald Denning, George Edmunds, Kenneth Hagen, Alan Hardy, Gentaro Imadaté, Dennis Kopp, Ronald McGinley, Woodrow Middlekauff, and Darwin Tiemann.

Betty Birdsall, Claudia Madison, and Elizabeth Randal read major portions of the manuscript for clarity of interpretation to non-biologists and consistency in grammar; and Arthur C. Smith reviewed a draft of the entire manuscript, making many useful suggestions regarding its content.

Line drawings accompanying the synopses of orders were prepared by Celeste Green (redrawn from Barker et al., 1977; Borror et al., 1976; Essig, 1942; and Hogue and Truxal, 1970).

Special acknowledgment is made to Liz, who helped JAP keep work in proper perspective, suffered through indescribably dreary tedium of written drafts, typing, proofreading, and corrections; she tolerated my exasperation with the discovery of each new error and the endless, delay-causing additions, and she helped me develop writing as well as other half-forgotten talents.

To all of these, the photographers listed below, and the many others who wittingly, willingly, or otherwise provided assistance with our education during the preparation of this book, we extend our grateful thanks.

The illustrations are original drawings, although many were made with reference to published photographs or drawings. This was done often with modifications of detail to clarify recognition features and to obtain likeness of natural postures of living insects. Acknowledgment is offered collectively for the contribution made by the original figures.

PHOTOGRAPH CREDITS

R. A. Arnold: 12e

G. Ballmer: 6f, 6g, cover (lower left)

Norman Bean: 13b

Jean Bourassa: 7h

L.R. Brown: 9b

James Carey: 7f, 12d

Patrick Craig: 3c, 3d, 3h, 5c, 5d, 7d, 9h, 13h, 15f

H.V. Daly: 8b, 9c

J.P. Donahue: 11c

E.B. Edney: 1g

D.K. Faulkner: 5e

Don Frack: 8d, 9a

R.W. Garrison: 1b

John Hafernik: 13f, 13g, 15c

K.S. Hagen: 15d

C.L. Hogue: 1a, 1e, 1f, 1h, 2a, 2c, 2d, 2e, 2f, 2g, 3a, 3b, 3e, 3f, 4c, 4d, 4f, 4g, 5a, 5b, 6a, 6b, 6d, 6e, 7b, 7c, 8a, 9f, 10d, 10e, 10f, 10g, 10h, 11a, 11b, 11d, 11e, 11g, 11h, 12a, 12b, 12c, 14a, 14c, 14d, 14e, 14h, 15b, 15g, 16c, 16d, 16g, 16h, cover (upper and lower right)

J. Hogue: 15h

F.T. Hovore: 13d, 14b, 14f, 15e

J. Levy and P. Bryant: 12c, 13e; and R. Vanderhof: 12f

J. Levy and E. Ling: 12h

Los Angeles County Museum: 4b, 5g, 10a, 10b

R.W. Merritt: 6c

R.J. Pence: 8g, 15a

J.A. Powell: 1c, 2b, 2h, 4a, 4e, 4h, 5f, 6h, 7a, 7e, 7g, 8c, 8e, 8f, 8h, 9d, 9e, 9g, 10c, 11f, 12g, 13a, 13c, 16a, 16b

Frank Skinner: cover (upper left)

Roy Snelling: 16e

S.L. Szerlip: 1d, 3g, 5h, 16f

Darwin Tiemann: 14g

NOTE: Figures on text illustration pages and on color plates are not necessarily reproduced to the same scale, even within a set shown on one page. Readers should refer to the text where sizes of the insects are given in the species accounts.

INTRODUCTION

Insects occur everywhere except in the polar regions and in the oceans. A few species even have adapted to life on the Antarctic shelf margins, cold, subarctic islands, in intertidal zones, and on the open ocean surface. Insects live in great abundance at every ecological horizon on land and in fresh water—in soil, where many feed on decaying organic matter or on plant roots; on primitive, low-growing plants such as mushrooms, mosses, and ferns; on or in the trunks of shrubs and trees; and on or in foliage, flowers, or seeds of flowering plants. At each of these horizons there is a complex of insects, each living in its own way, as a scavenger, primary consumer of plant material, predator, or parasite of other insects or other animals, and so on. Among both plant feeders and parasites, there are some general feeders and some that show great specialization for particular hosts.

GROWTH AND REPRODUCTION

In most insects males and females are produced in about equal numbers, mating occurs, and females deposit eggs on or near the food used by the young. There are many modifications of this basic pattern—males may be produced only at certain seasons, as in social insects; generations of females may be produced without mating (parthenogenesis) while environmental conditions are favorable, as in aphids; or sexual forms are lost altogether, a phenomenon known in certain weevils, moths, and other insects. In normal forms the sexes locate one another extremely effectively, and most females are mated almost as soon as they emerge. Commonly, attraction of mates is accomplished by means of a volatile chemical or scent (pheromone) broadcast in the air, usually by the female. In some insects, such as butterflies and fireflies, visual perception of the mate is necessary; in others, like crickets and cicadas, calls or sounds are perceived by one or both mates.

Adult insects do not grow. Growth occurs through a series of stages (instars); at the end of each, the shell or skin is shed, and the insect enlarges. After the final instar, molting produces the adult stage. In the simplest life cycle, such as that of silverfish, roaches, and grasshoppers, the eggs hatch into tiny nymphs resembling the adults except that they are smaller and lack wings and reproductive organs. Most insects have a more complex life cycle, normally with 4 stages (holometabolous—a complete change or metamorphosis): adults lay *eggs* which hatch into caterpillars, maggots, or various other forms of *larvae* which feed and grow through a series of molts. A non-motile *pupa* is produced at the last larval molt. A cocoon of silk may or may not be formed by the larva when it is ready to form the pupa, varying with the kind of insect. Finally, transformation to the *adult* occurs within the pupa, the shell of which is broken by emergence of the adult.

An intermediate kind of change or metamorphosis (hemimetabolous) occurs in certain aquatic insects, such as dragonflies and mayflies. The young stages, called nymphs or naiads, are quite different from the adults which do not live in water, and the transformation takes place in a motile but non-feeding stage (''pre-imago'').

In an evolutionary sense the adaptation to a complete metamorphosis was perhaps the single greatest achievement by insects. It enabled them to exploit a vast range of habitats by allowing adults and larvae to occupy different niches, and it allowed more versatile development of a resting stage, in which all growth and development stops (diapause). This resting stage enables insects to synchronize activities to favorable seasons and to enhance survival during unfavorable times, such as the long dry season in California, by remaining dormant. Diapause may occur in the egg, larva, pupa, or adult, varying with the species. Many insects can even wait two or more years to complete the life cycle if seasonal conditions are highly unfavorable or uncertain, as in desert habitats.

BREATHING AND CIRCULATION

Insects and related animals, such as sowbugs, spiders, and lobsters (Arthropods), have an exterior shell or skeleton within which the blood freely circulates. In insects the blood is moved by action of a tubular heart located in the upper part of the body cavity, opening just behind the brain. Blood is pumped forward to the head area, from which it circulates to the organs and into the appendages before reentering the posterior end of the heart.

In the majority of insects, air is taken in through holes in the body wall (spiracles). These are located along the sides of the body, usually one pair to each segment, 2 on the thorax, 8 on the abdomen in the adults, and fewer in larvae. Tubes (tracheae) lead from the spiracles throughout the body cavity. These ramify through the organs and appendages, branching into finer and finer tubes (tracheoles). Oxygen passes from the tracheoles into the body, and carbon dioxide diffuses in the reverse direction and through the body wall.

Some aquatic insects breathe in the usual manner, by periodically coming to the surface to collect a bubble of air. In many aquatic forms, however, breathing occurs through the general body surface or through special organs or gills. The latter are thin-walled outgrowths located along the sides of the thoracic and/or abdominal segments or at the tail end. Oxygen diffuses directly from the water through the gill walls, which are well supplied with tracheae and tracheoles. Tracheal gills are present in immature stages of mayflies, dragonflies, damselflies, stoneflies, caddisflies, alderflies, and many aquatic two-winged flies. Therefore these forms are able to spend their whole lives beneath the surface of the water.

FEEDING

Diverse methods of feeding are employed. The kind of mouthparts and the mode of feeding are generally correlated with major groups or orders of insects (see "Structure and Classification," below). Thus, grasshoppers and related insects have "jaws" or mandibles adapted for biting and chewing, as do beetles, wasps, and bees, and most other insects. By powerful musculature, the horizontally opposing mandibles are capable of biting through solid food substances, which may be as tough as hard wood. In other insects, such as aphids, leafhoppers, true bugs, and some flies, the mouthparts are formed into a tubular beak which is stabbed into the food, and juices are sucked up by means of a pumping action. Butterflies and moths have the mouthparts modified into a long, coiled tube which is inserted into food sources such as nectar in flowers, and this is taken up by a pumping action. In some insects enzymes are ejected to liquify a food source, such as dried honeydew, before it can be sucked up. In some cases, as in mosquitoes and fleas, a salivary fluid is injected which prevents coagulation of mammalian blood. Many kinds of flies are adapted to feed on blood, and they have piercing and sucking mouthparts, while other flies take up their food via a proboscis with a spongelike tip.

Most insects are harmless to man from a medical or health standpoint. Lice, fleas, bedbugs, and a few other true bugs, mosquitoes, and certain other flies are familiar for their bites. Some bugs and flies, normally predaceous on other insects, can inflict a painful bite if prompted, but none of these injects a poisonous toxin comparable to that of a black widow spider. On the other hand, certain diseases are transmitted by insects, especially in tropical regions, commonly by biting. In California these include plague, transmitted by fleas from rodents; encephalitis, sometimes called "sleeping sickness," from birds by mosquitoes; and malaria, carried by mosquitoes. None of these is epidemic today, although a few cases of encephalitis and plague are recorded each year.

STINGING

Few insects are capable of stinging, but often the term "sting" is misused for insect bites. The sting apparatus is a modified ovipositor in female ants, wasps, and bees. No other insects sting. In contrast to insect bites, the sting is accompanied by the injection of a toxin, a mixture of complex proteins and enzymes which acts on the tissues of the victim to release histamine; and in humans, depending upon the individual sensitivity, the effects may be severe. Many kinds of wasps and bees are capable of stinging if provoked, but the behavior has its most damaging effect in social species, ants, yellowjackets, bumble bees, and the Honey Bee, where the workers rush out in great numbers to attack intruders, activated by alarm chemicals given off by their sisters. Some ants both sting and bite. In the Honey Bee worker, the sting is equipped with barbs so that it fixes in place and is pulled out of the bee, which kills her. Workers are expendable for the protection of the colony.

DIVERSITY AND DISTRIBUTION OF THE CALIFORNIA INSECT FAUNA

California's size, long latitudinal range, and topographical features set the stage for tremendous biological diversity. The second largest of the 48 contiguous states, comprising an area about equal to the New England states, Pennsylvania, and New York combined, California by size alone would be expected to have large numbers of plant and animal species. In addition, several geographic and climatic zones are created by the topography, which is dominated by two mountain ranges. These ranges parallel the coast and serve to modify the ocean's influence. Elevations range from above 4200 m (14,000 ft) in the Cascade Range and Sierra Nevada to below sea level in the Imperial and Death valleys. Annual rainfall varies from more than 270 cm (110 in) in Del Norte County to less than 5 cm (2 in) in parts of the desert. In most parts of the state nearly all the precipitation occurs between October and May. Coastal areas often have 365 frost-free days per year, while there are fewer than 100 at high elevations. Fog is also an important climatic influence, particularly along the coast where it is more frequent in summer, and it may be an important winter influence in the interior valleys.

TOPOGRAPHY

The major topographical features (Map 2) of the state consist of two major mountain chains, the Coast Ranges and the Sierra Nevada, which are joined in the north by the southern end of the Cascade Range and in the south by the Transverse Range. The Sierra Nevada is the dominant feature; it consists

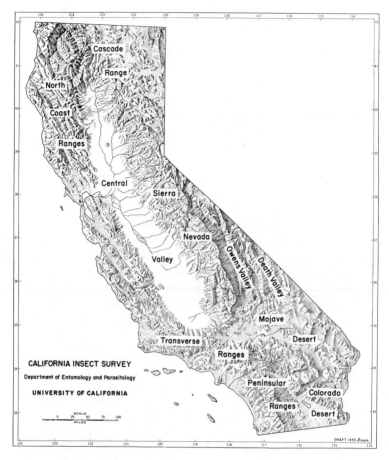

Map 2 Major topographical features of California.

of an immense granitic uplift more than 600 km (360 mi) long, extending from Plumas County to Kern County. In general its east side rises abruptly from 1500 to 3000 m (5,000–10,000 ft) above the basins to the east and descends gradually to the west via a tipped plateau, divided by a series of deep canyons of the west-flowing rivers. The Coast Ranges, subdivided into outer, marine-influenced, and inner, drier ranges, are lower but reach elevations above 2500 m (8,000 ft) in the north. The Transverse and Peninsular ranges also reach elevations of 2500 m and 3000 m (8,000–10,000 ft), and, like the others, serve to block the ocean's influence, creating a "rainshadow" effect and aridity in interior regions.

The topographical features combine to create a series of parallel zones, each with its characteristic plant and animal communities. These zones are mostly distributed in north-south, linear patterns. Proceeding inland, these are (1) coastal strand and coastal plain; (2) Coast Range foothills; (3) Coast Range conifer zone; (4) Central Valley; (5) Sierra Nevada foothills; (6) Sierra Nevada conifer zone; (7) high alpine Sierra Nevada; and (8) the arid interior. In southern California, the Peninsular Range zones give way abruptly to the low Colorado Desert (0–500 m; 0–1500 ft). East of the Transverse Ranges and southern Sierra Nevada lies the higher Mojave Desert (600–1000 m; 2000–3,000 ft) with its own mountain ranges; further north the Sierra Nevada and Cascade ranges impose their influence on the Great Basin, which is dry sagebrush country to the east and mostly above 1000 m (3000 ft) in elevation.

DIVERSITY

Summer visitors from the eastern United States, especially if their travels are restricted to lowland areas, around the cities, and major north-south highways, often remark on the dryness of the California countryside. Insect collectors from places like Minnesota or Illinois, arriving here in June when the collecting season is just getting into full swing there, are dismayed to find that it is about finished in lowland California. Unless they obtain local advice or plan to work in the mountains, they are likely to find summer collecting in California a bleak prospect; and they may justifiably find it hard to believe that we have a larger number of insect species than nearly any other state.

The mild climate with a long dry season is a key to the hidden diversity. Prolonged drought imposes a strong seasonal partitioning on plants and animals. In California this is superimposed on the topographical diversity, creating a complex mosaic of geographical, ecological, and seasonal niches to which insects have adapted. Northern, higher elevation, and interior regions of North America have a short growing season, with the vast majority of the insects active during the summer months. In most of California, by contrast, the growing season for annual plants takes place mainly during winter and spring. The insects feeding on annuals and using their pollen, as well as associated parasitic insects, must complete their activity early, often ahead of the growing and flowering seasons of perennial plants. For example, small moths are active in January and February in southern California and along the immediate coast northward. They are dayfliers and are dark colored to take advantage of the sun's warming at a season when temperatures at night are too cold for activity. This early start enables their resulting larvae to complete feeding before the annual plants on which they specialize dry by late March or April. Comparable adaptations occur in large numbers of other insects. Even species associated with broadleafed shrubs and trees often complete their activity early. Leafing out generally occurs at the end of the rainy season, in April and May, and the leaves harden, becoming more resistant, by summertime. Many other insects are active in fall, such as those associated with fall-blooming composite shrubs. Many fall-fliers feed as larvae in spring, overwintering in the egg stage.

Therefore, in order to study the insects of California, all seasons have to be taken into account. At lower elevations there are many spring-fliers in January to April, but there are summer species in marshes, river habitats, and in association with perennial plants. Elsewhere in the state, insect activity peaks in the foothills in May, at intermediate elevations in June, and in the high Sierra in July or August. By the time regular frosts occur at timberline, fall-flying species are active at low elevations. Even during the most dormant period, mid-November to mid-January, there are insects to be obtained. This is the time adults of many species, including some particularly unusual kinds, reach peak abundance, especially in the deserts and at coastal localities.

Whereas 50 species of a given family may be collected by sampling one or a few localities in Wisconsin or Ohio (possibly the same 50 in both states), a comparable number may be known in California, but the beleaguered collector must work in a dozen habitats from February to August to find them. It is no wonder that most habitats have not been adequately sampled and that discovery of new species in California is commonplace. Occasionally we rediscover a species known only from specimens taken by an early explorer and not collected for the past 100 years or more.

GEOGRAPHICAL DISTRIBUTION

Various methods have been developed for analyzing the distribution of animal species. These include the defining of vegetation types, life zones, and faunal provinces.

Botanists have classified the flora of California by plant communities. Most of these have a climatic basis, but some are highly influenced by soil type. They vary in distinctiveness and in complexity of distribution pattern. About 30 plant communities are defined by Munz and Keck, in *A California Flora,* ranging from Coastal Strand and Redwood Forest to Alpine Fell-fields and Pinyon-Juniper Woodland. The plant communities are in turn grouped into 11 *vegetation types,* which generally are more useful than the individual plant communities in describing insect distributions. A great many insect species, especially plant-feeding forms, are restricted to one of the vegetation types; (1) Coastal Strand; (2) Salt Marsh; (3) Freshwater Marsh; (4) Scrub; (5) Coniferous Forest; (6) Mixed Evergreen Forest; (7) Woodland Savannah; (8) Chaparral; (9) Grassland; (10) Alpine Fell-fields; (11) Desert Woodland.

The Merriam Life Zone concept, proposed before the turn of the century for agricultural prediction, is based on temperature criteria. In general more northern and higher elevation areas are colder and have shorter growing seasons. Applied to the midwestern United States, broad belts representing successive north to south zones are produced. In the west, the diverse topography results in a series of narrow zones flanking the mountains, and these zones sometimes are useful in interpreting and predicting distributions of animals, including insects. Specific zones are recognized by a few indicator species, especially trees, and are therefore somewhat arbitrarily delimited. In general, insects are more broadly distributed than a single life zone, and use of individual zones is more specific than our needs. In California we can consider the life zones in two groups: (1) the lower or *Austral zones* (Lower Sonoran and Upper Sonoran), encompassing the Central Valley, interior valleys, and deserts, and the arid south coastal plain; and (2) the *Boreal zones* (Transition, Canadian, Hudsonian, and Arctic Alpine), corresponding to portions of the state characterized by pine forests.

Perhaps the best approach to analysis of insect distribution is one in which faunal provinces are proposed, based on a combination of topographical features, life zones, and ecological or vegetational types. Many attempts have been made by biologists to define useful biotic provinces, but none has unanimous acceptance, primarily because various plants and animals differ in adherence to one or another of the criteria used to create the provinces. One of the best analyses for California is that of Alden Miller, who proposed six broad provinces on the basis of the distribution of birds. Some of the provinces were subdivided into several more specific districts. Because most insects are more widely distributed than a given life zone or plant community, larger provinces provide convenient categories for geographical distribution analysis. Map 3 gives a modified version of Miller's provinces, revised on the basis of data from the *Bulletin of the California Insect Survey.*

Six provinces are defined, each including several vegetation types. Perhaps half the state's insect species are limited to one or a portion of one of these provinces, including the majority of plant-feeding types. Many species, including general feeders such as scavengers and predators, are more widespread. Although parasitoid insects often are specific to their host or a particular habitat, many are less restricted and are widely distributed. There are also many plant-feeding species which either feed on a wide variety of plants (polyphagous) or are restricted to one plant family or genus but use representatives

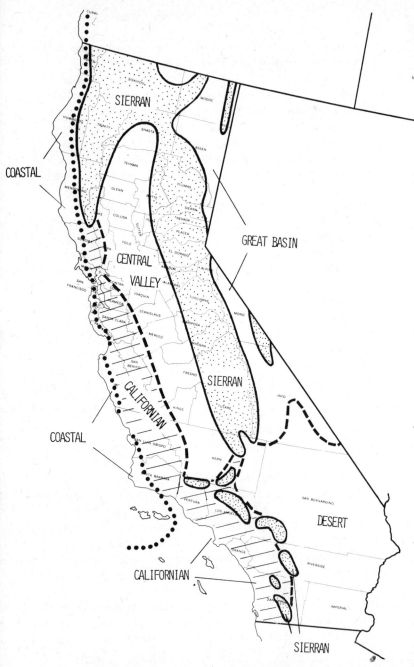

Map 3 Major insect faunal provinces of California.

that grow in diverse ecological situations. The entomofaunal provinces are as follows:

1. *Coastal*. Characterized by plants and animals of northern affinities and including parts of the coastal strand, coastal scrub, and northern coastal coniferous forest, this province has a high proportion of insects that range along the coast from Vancouver Island or further north. There are few species known only from California (endemic) in this area.

2. *Sierran*. Containing most of the coniferous forests of the state and including a range from 450–3000 m (1500–10,000 ft) elevation, probably this is the most diverse in habitats and the richest province in insect species. Owing to its diversity, the Sierran province has a high proportion of species that range into adjacent zones, such as the foothill woodland of the Central Valley and the Coast Ranges. Nevertheless there are vast numbers of insects found only in this province. In general these are not restricted to California.

3. *Central Valley*. Considered a portion of the Californian Province by Miller, this area shares an appreciable number of its insects with the Mojave Desert, especially those of the western San Joaquin Valley, which has sandy areas harboring desert plants. Many other insects occur widely in the Californian Province.

4. *Californian*. So named because it has a larger proportion of species endemic to California than any other province, this area is also diversified. It contains the southern coastal strand and coastal scrub as well as the chaparral, foothill woodland, and coniferous forests of the southern ranges. All of these are relatively arid compared to their northern counterparts. Although most insects range northward into the Coastal or Sierran province, a large number occur only here and in adjacent northern Baja California.

5. *Great Basin*. This area represents the western edge of vast sagebrush scrub areas to the east and north of California. It is rather uniform in character, rich in species, and relatively distinct from other parts of the state's fauna. It contains almost no insects endemic to California.

6. *Desert*. There are many differences between the low Colorado Desert and higher Mojave, but an overall similarity

in insect species predominates. The area has a strong distinctness within California but few endemics. The vast majority of species which are restricted to this province in California range into Arizona and beyond.

The Channel Islands were classified as a distinct province by Miller, but the islands' insects generally are those of the mainland. The insects are not well inventoried, but it appears that most of those of the northern island group, especially Santa Cruz Island, are species of the Californian and Coastal provinces, with few endemic species. By contrast the southern islands, especially the more distant ones, have higher numbers of endemics and a greater affinity with the southern part of the Californian and the Desert provinces.

MICROHABITATS

Within any one of the life zones, vegetation types, or faunal provinces there are a large number of different habitats varying with the season. Habitats may be large or small in extent, but each has only a narrow range of environmental conditions. The following are some examples of microhabitat types, and each may be further subdivided according to shade, moisture conditions, etc. because insects specialize by moisture tolerance or other factors.

List of Twenty Common Microhabitats with Characteristic Insect Examples

1. On ground under stones or debris (webspinners, predaceous ground beetles, darkling ground beetles).
2. In sand of beach or desert dunes (burrowing bugs, ground beetles, weevils, larvae of moths).
3. In moist leaf litter (springtails, diplurans, many kinds of small beetles).
4. In freshwater ponds (diving beetles, backswimmers, water boatmen, naiads of damselflies and dragonflies).
5. On or under stones in swift-moving streams or cascades (immatures of stoneflies, caddisflies, water pennies, and black flies).
6. On surface of pools and ponds (water striders, riffle bugs, whirligig beetles).

7. On sand or mud margins of streams or ponds (toad bugs, shore bugs, tiger beetles).

8. In galls of woody plants (larvae of gall wasps, gall gnats, or small moths, and parasitic wasps of all 3).

9. On or in leaves of living plants (aphids, whiteflies, lacebugs, leafminer moth and fly larvae).

10. Visiting flowers (various beetles, hoverflies, butterflies, bees).

11. Inside fruit or seed of flowering plants (larvae of moths, weevils, and fruit flies).

12. On or in wood rot fungi (fungus gnats, moth larvae, many kinds of small beetles).

13. On recently felled tree branches or trunks (bark beetles, metallic wood borer beetles, longhorned beetles, parasitoid wasps).

14. Under bark of old stumps or logs (ironclad beetles, carpenter ants, various fly larvae).

15. In punky wood of old rotting logs (termites, adults and larvae of click beetles and stag beetles).

16. On recently killed vertebrate animal carcasses (blowflies, hister beetles, carrion beetles).

17. On old, drying vertebrate carcasses (hide beetles, rove beetles, tineid moth larvae).

18. On or in animal droppings, especially cow pats (dung flies, blow flies, scarab beetles).

19. On living mammals or in their fur (horse flies, fleas, lice).

20. On human skin (head lice, crab lice).

STRUCTURE AND CLASSIFICATION

Animals are classified into a hierarchy of categories called *taxa* (singular = taxon) with each higher taxon containing one to many taxa of the next subordinate rank. Following are the levels commonly used for insects, in descending order:

Phylum
　Class
　　Order
　　　Suborder
　　　　Superfamily
　　　　　Family
　　　　　　Subfamily
　　　　　　　Genus
　　　　　　　　Species

The arrangement is based on many characteristics in insects, including features of adults and immatures, and is presumed to reflect natural or evolutionary relationships. Thus, predictions about species beyond the information already available can be made, based on what is known about closely related species. For example, a wasp of the genus *Ammophila* (p. 342) may be known only from adult specimens, yet it is possible to predict with confidence that the species preys on moth caterpillars, digs burrows of a certain type, supplies caterpillar provisions to one offspring per burrow, and so on, because all other *Ammophila* for which habits are known do so. It would be possible to classify animals in some artificial way, such as by numbers or alphabetically by name, which would be useful in locating and retrieving information in a library or museum, but the predictive quality would be lost.

Insects and other animals with an outside shell for a skeleton and jointed legs, including spiders, lobsters, sowbugs, etc., make up the *Phylum Arthropoda*. Insects are distinguished from other arthropods by having three body regions, head, thorax, and abdomen, and by having only three pairs of legs. Within the *Class Insecta* there are about 30 *orders,* their number varying with authorities' interpretations; the major orders are characterized below. For example, beetles make up the Coleoptera, the largest order in numbers of species.

Because there are so many kinds of insects, it is impossible to know and talk about all the species even for a local area. Therefore the *family* is most often the level at which communication about insects occurs. Most families contain species of generally similar appearance and habits. For example, beetles of the family Cerambycidae (p. 301) are generally recognizable by their long antennae, and their larvae are all similar-appearing grubs which bore in roots or wood of living or dead trees and shrubs. When students learn to recognize 300 or 400 of the common or distinctive families in a region such as California, they can communicate about the vast majority of insects around them. Moreover, much of this information is applicable in many other parts of the world because the major families are worldwide in distribution. Studies of living primitive peoples have shown that communication systems about entities such as plants or animals generally contain about 250–800 items, those for which recognition and the ability to communicate names are useful to the people. This helps to show us why we cannot communicate information easily about 28,000 species of insects in California, but a few hundred dominant families fall within a range most people can comprehend.

NAMES

The *scientific names* of animals are based on the two lowest rank categories, the *genus* (plural = genera) and the *species* (plural and singular = species). Therefore, each insect has two parts to its name: the generic name, the first letter of which is capitalized, followed by the specific name, which is not capitalized. For example, the scientific name for the Mourning Cloak butterfly is *Nymphalis antiopa* (p. 250), while the California Tortoise Shell butterfly, a member of the same genus, is *Nymphalis californica* (p. 249). The names are latinized and are usually printed in italics. In technical literature and in a properly labeled collection, the name of the author who first named and described the species is also given; these are included here only in the index. The simplicity of this system is so attractive that zoologists have used it universally ever since it was proposed by Linnaeus in the mid-1700s. In what are perhaps the only international agreements accepted by all nations, zoologists and botanists throughout the world use sets of rules governing the names of animals and plants. Every animal has a unique name. The generic name of an insect cannot be used for any other kind of animal, but a specific name can be used more than once, so long as the species are in different genera. For example, many species in different genera are named *californica.*

In many cases industrious taxonomists have added a third name, the subspecies, to describe geographically limited populations that differ in color or other respects. In general we do not mention these, even though they have been applied to many of the species we discuss.

There are no rules regarding *common names,* and they change from place to place for the same animal. For instance, *Nymphalis antiopa* is called the Mourning Cloak here, but in England it is the Camberwell Beauty. Although there is a list of common names approved by the Entomological Society of America for the more important species in agriculture, laboratory research, and so on, most insects have never had common names proposed. We have adopted the scientific and common names given in that list with few exceptions. In cases where we were unable to locate accepted common names, new ones are applied in this book.

SYNOPSIS OF THE ORDERS OF INSECTS

For convenience, the orders of insects are organized into two major groups, APTERYGOTA, the primitive wingless insects, and the PTERYGOTA, the winged insects. These groups do not represent formal taxa of comparable rank to one another because the Apterygotans consist of unrelated orders, in part considered to make up another class outside the Insecta by many writers. Some kinds of Pterygota have become flightless through secondary loss or reduction of wings, as in lice, fleas, ants, and occasional members of other groups. The Pterygota have two major subdivisions: the hemimetabolous insects, those with a partial change in form through the life cycle, and the holometabolous groups, those with the full metamorphosis through egg, larva, pupa to adult. Again, these are not formal taxa of comparable rank. Hemimetabolous insects (orders Ephemeroptera through Thysanoptera in our list) include groups with various kinds of incomplete metamorphosis. The holometabolous orders (Neuroptera through Hymenoptera) include about 85 percent of living insect species, yet all have a similar, complete metamorphosis.

PRIMITIVE WINGLESS INSECTS
Apterygota

PROTURANS
Order Protura

Minute, slender, delicate. Head pear-shaped; eyes, ocelli, and antennae absent; mouthparts formed for sucking, retracted within head. Legs short, tarsi with only one segment, each bearing a single claw; wings absent. Abdomen composed of eleven segments, the last three very short, the basal three with ventral, paired styli (vestigial legs); cerci absent. Metamorphosis gradual; immatures similar to adults except for size; molting continues throughout life with segments added terminally to abdomen.

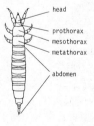

head
prothorax
mesothorax
metathorax
abdomen

DIPLURANS
Order Diplura

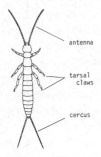

Small to medium-sized, slender, unpigmented. Head large, oval; antennae long, with numerous beadlike segments; eyes and ocelli absent; mouthparts mandibulate, retracted within head. Thoracic segments similar; legs similar, tarsi with only one segment, each bearing 2 claws; wings absent. Abdomen with ten segments, some with ventral, paired styli (vestigial legs); cerci well developed, either (a) long, filamentous, and many segmented, (b) short and few segmented, or (c) unjointed and forcepslike. Metamorphosis gradual; immatures similar to adults except for size.

SPRINGTAILS
Order Collembola

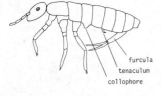

Minute to small, soft-bodied, cuticle (outer layer of body wall) often with scales, rods, tubercles, etc. Compound eyes absent or represented by groups of separate facets; antennae short with few segments; mouthparts chewing or sucking, retracted within head. Abdomen with only six segments, some bearing unique ventral appendages: median projection (collophore) on segment 1, elongate springing apparatus (furcula) on segment 4 or 5, and catch (tenaculum) for holding the furcula on segment 3; cerci absent. Metamorphosis gradual; immatures similar to adults except for size.

SILVERFISH AND BRISTLETAILS
Order Thysanura

Small to moderate-sized, often clothed with scales. Antennae long, hairlike (filiform), with many segments; compound eyes present, large and dorsal or small and lateral; ocelli present or absent; mouthparts mandibulate, exserted. Legs similar, tarsi with two or three segments, bearing 2 claws. Abdomen with eleven segments, some with large, ventral, paired styli (vestigial legs); long, filamentous cerci and similar median structure present. Metamorphosis gradual; immatures similar to adults except for size.

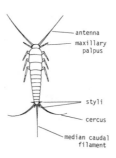

WINGED INSECTS
Pterygota

MAYFLIES
Order Ephemeroptera

Small to moderate-sized, delicate. Compound eyes large (sometimes secondarily divided); ocelli present; antennae with two large basal segments, the remainder threadlike (flagellum); mouthparts vestigial. Prothorax small, free, mesothorax large; wings present, 2 pairs, membranous, fore pair much larger than hind, with complex net venation.

Abdomen slender, tipped with 2 or 3 long, many-segmented filaments. Metamorphosis gradual; nymphs (naiads) aquatic, with abdominal gills; winged immature form (sub-imago) precedes final molt to adult.

DRAGONFLIES AND DAMSELFLIES
Order Odonata

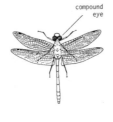

compound eye

Medium-sized to large, slender. Eyes very large with units (om-matidia), extraordinarily numerous; ocelli present; antennae minute; mouthparts biting, mandibles strongly developed. Thoracic segments fused and tilted obliquely backward; legs similar, placed far forward, armed with heavy spines; wings present, 2 pairs, similar, membranous, complexly veined. Abdomen long and narrow; cerci 1-segmented; secondary male copulatory apparatus attached ventrally to second and third segments. Metamorphosis gradual; nymphs (naiads) aquatic with gills, change directly to adults.

STONEFLIES
Order Plecoptera

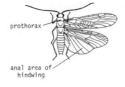

prothorax

anal area of hindwing

Small to large, soft-bodied. Head broad and flattened; compound eyes and ocelli present; antennae long, many segmented; mouthparts man-dibulate, sometimes reduced. Prothorax large, free; 4 wings, membranous hind pair larger than fore, with large posterior (anal) area, both pairs with complex venation. Metamorphosis gradual; nymphs (naiads) aquatic, with filamentous gills, change directly to adults.

COCKROACHES
Order Blattodea

Medium-sized, flattened, with the head nearly or completely covered from above by large, shieldlike pronotum. Antennae threadlike (filiform) with numerous segments; mouthparts mandibulate. Legs similar to one another; tarsi 5-segmented. Forewings thickened, somewhat leathery, held flat over back; hindwings broader, membranous, fan-shaped, folding under forewings when at rest. Abdomen with 2 short, many-segmented cerci. Nymphs similar, lacking wings. Metamorphosis gradual; nymphs resemble adults except lacking wings.

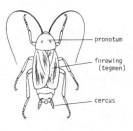

TERMITES
Order Isoptera

Small to medium-sized, elongate, polymorphic (castes) with highly developed social habits. Body usually soft, eyes usually present in reproductive castes, absent in others; mouthparts chewing, for feeding on wood (worker caste only, very large mandibles used for defense in some soldier castes); prothorax free;

wings usually absent, when present, only temporary (in dispersing adults), similar, with strong anterior veins only, shed after new colonies founded. Metamorphosis gradual; immatures similar to adult workers in structure and habits.

ROCK CRAWLERS
Order Grylloblattodea

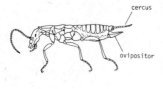

cercus

ovipositor

Wingless, elongate, with eyes reduced or absent. Antennae moderately long, threadlike; mouthparts mandibulate. Legs similar to one another, tarsi 5-segmented. Female with large ovipositor. Abdominal cerci long, 8-segmented. Metamorphosis gradual; nymphs similar to adults.

GRASSHOPPERS, CRICKETS, KATYDIDS
Order Orthoptera

Mostly medium-sized to large, winged or wingless. Mouthparts mandibulate; antennae many segmented, short, thick or long, threadlike. Hind legs usually enlarged for jumping. Forewings narrow, usually thickened, leathery, folded flat or rooflike over back; hindwings membranous, fanshaped, often brightly colored, folded under forewings when at rest. Abdominal cerci short and unsegmented. Specialized hearing and sound-producing organs frequently developed. Metamorphosis gradual; nymphs similar to adults except lacking wings.

MANTIDS
Order Mantodea

Medium-sized to large, winged. Antennae threadlike with numerous segments; mouthparts mandibulate. Forelegs fitted for grasping prey (raptorial); mid- and hindlegs unmodified; tarsi 5-segmented. Forewings thickened, leathery, often short; hindwings fan-shaped, often colorful, folded under forewings when at rest. Abdominal cerci many-segmented. No hearing or sound-producing organs. Metamorphosis gradual; nymphs similar to adults except wingless.

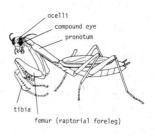

WALKING STICKS
Order Phasmatodea

Mostly large, winged or wingless, but sticklike and wingless in all our species. Mouthparts mandibulate; antennae many segmented, short or long, and threadlike. Prothorax short, meso- and metathorax elongate; legs widely spaced, similar to one another; tarsi usually 5-segmented. Cerci short, unsegmented. No hearing or sound-producing organs. Metamorphosis gradual; nymphs similar to adults.

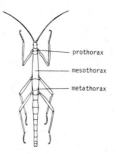

EARWIGS
Order Dermaptera

Small to moderate-sized, elongate, heavily sclerotized. Compound eyes and ocelli well developed or absent; antennae threadlike (filiform), many segmented;

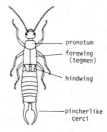

mouthparts biting. Thoracic segments distinct; legs short, heavy, tarsi 3-segmented; wings usually present, 4, fore pair leathery, short, leaving abdomen exposed, hind pair large, broadly oval, membranous, with many radiating veins, folding under forewings when at rest. Abdomen usually with heavy, 1-segmented, pincherlike cerci. Metamorphosis gradual; nymphs similar to adults in structure and habits.

WEBSPINNERS
Order Embioptera

basal segment
of tarsus

Small to medium-sized, soft-bodied, elongate. Eyes present; ocelli absent; antennae long, many segmented; mouthparts biting. Prothorax large, free; first tarsal segments of forelegs containing silk glands for spinning webs in which the insects live; wings absent in females, usually present in males, 2 pairs, similar, elongate, venation simple. Abdomen with short cerci, tip of abdomen in male usually different on the two sides (asymmetrical). Metamorphosis gradual; nymphs resemble adults in structure and habits.

BARK LICE
Order Psocoptera

Minute to small, soft-bodied. Head and eyes prominent, mouthparts chewing; prothorax usually small, meso- and metathorax usually separated; wings absent or present, membranous, with simple venation, few crossveins, fore pair larger than hind, sometimes scaly or hairy; legs slender, tarsi with two or three segments; cerci absent. Metamorphosis gradual; immatures similar to wingless adults in structure and habits.

BITING LICE
Order Mallophaga

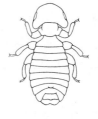

Minute to small, wingless ectoparasites primarily of birds (some on mammals). Body strongly flattened; antennae simple, short, exposed or lying in ventro-lateral grooves; mouthparts biting, ventral, for feeding on feathers, fur, or skin; prothorax free, rarely fused to mesothorax; legs tipped with normal claws, no special grasping mechanism for clinging to feathers or hairs; no cerci. Metamorphosis gradual; nymphs similar to adults in structure and habits.

SUCKING LICE
Order Anoplura

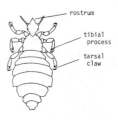

Minute to small, wingless ectoparasites of mammals. Body flattened; eyes reduced or absent; antennae simple, short; mouthparts suctorial, long slender piercing stylets opening on short, anterior, unjointed rostrum, adapted for feeding on vertebrate blood; thoracic segments fused; legs stout, tibial process opposing tarsal claw to form a grasping mechanism for clinging to hairs; no cerci. Metamorphosis gradual; nymphs similar to adults in structure and habits.

TRUE BUGS, HOPPERS, APHIDS, SCALES
Order Hemiptera

Minute to large. Mouthparts fitted for piercing and sucking, mandibles and maxillae present as long, slender stylets enclosed in a beaklike labium, projecting downward (Heteroptera) or backward (Homoptera). Two pairs of wings usually present, forewing (hemelytron) thick at base and membranous apically, hindwing membranous entirely (Heteroptera) or both wings membranous (Homoptera). Metamorphosis gradual; nymphs similar to adults in structure and habits.

THRIPS
Order Thysanoptera

Minute to small, slender. Mouthparts fitted for sucking, consisting of a posteriorly directed, cone-shaped sheath that encloses the mandibular and maxillary stylets. Two pairs of wings usually present, similar, narrow, with few or no veins, fringed with long hairs. Tarsi with eversible bulb at tip. Metamorphosis gradual; nymphs similar

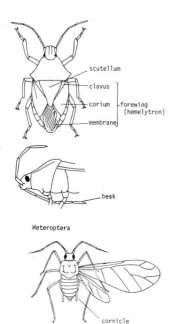

scutellum

clavus

corium — forewing (hemelytron)

membrane

beak

Heteroptera

cornicle

Homoptera

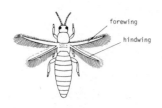

forewing

hindwing

to adults in structure, except wing-less; terminal instars often quiescent.

NERVE-WINGED INSECTS
Order Neuroptera

Small to large, soft-bodied. Mouthparts mandibulate; antennae usually long, threadlike, with many segments; wings always present, 2 pairs, similar, membranous, naked or moderately hairy, venation generalized to complex, with many accessory veins, often forked at wing margin, costal cell with numerous crossveins. Metamorphosis complete; larvae silverfish-shaped (campodeiform), with biting or suctorial mouthparts, mostly terrestrial, some aquatic; pupae free or in cocoons, active.

SCORPIONFLIES
Order Mecoptera

Small to medium-sized, slender. Head nearly always prolonged, beaklike, with well-developed mandibulate mouthparts at the tip; antennae long, threadlike; prothorax small, free; meso- and metathorax similar; wings absent or present, 2 pairs, membranous, usually long, narrow, and naked, with few crossveins; legs long and slender; genitalia of male often enlarged and forming a reflexed bulb. Metamorphosis complete; larvae caterpillarlike; pupae free, active.

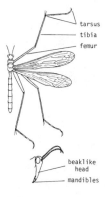

tarsus
tibia
femur

beaklike head
mandibles

GNATS, MIDGES, FLIES
Order Diptera

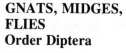
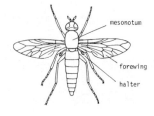

mesonotum

forewing

halter

Minute to moderate-sized, rarely large, usually with exceptionally good powers of flight. Mouthparts suctorial, usually constructed for lapping or piercing. Both prothorax and metathorax small and fused with the prominent mesothorax. Only the mesothoracic wings developed, the veins and crossveins not numerous; hindwings replaced by small, knobbed structures (halteres). Metamorphosis complete; larvae almost always legless maggots or grubs, never with true jointed legs, frequently with indistinct head and retracted mouthparts; pupae either free or encased in a capsule (puparium) formed from hardened larval exuviae.

FLEAS
Order Siphonaptera

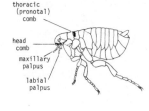

thoracic
(pronotal)
comb

head
comb

maxillary
palpus

labial
palpus

Small, wingless, strongly compressed side to side, jumping, with dark colored, hardened (sclerotized), bristly body and legs. Ectoparasitic as adults on mammals or rarely on birds. Mouthparts fitted for piercing and sucking; eyes present or absent; antennae short and stout, reposing in grooves; thoracic segments free; coxae large. Metamorphosis complete; larvae elongate, cylindrical, legless with well-developed head and biting mouthparts; free-living; pupae enclosed in cocoons.

CADDISFLIES
Order Trichoptera

Small to medium-sized, slender, mothlike. Antennae long, threadlike; mouthparts atrophied except for palpi; prothorax small, free; meso- and metathorax similar, wings similar, membranous, with few crossveins, densely clothed with hairs; legs long with tibial spurs. Metamorphosis complete; larvae aquatic, usually with tufted gills, either free-living or in cases constructed of silk and sand or plant particles or in silken webs or nets; pupae exarate, with functional mandibles.

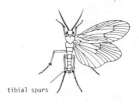

tibial spurs

MOTHS AND BUTTERFLIES
Order Lepidoptera

Small to large. Wings and body clothed with modified, flat hairs (scales) which usually form a color pattern. Antennae long, variously modified, threadlike, comblike (pectinate), or clubbed; mouthparts reduced or suctorial, composed of a long tube coiled under the head when not in use (proboscis); 4 wings, the 2 pairs similar in texture but not in shape, venation moderately complex with few crossveins.

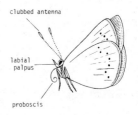

clubbed antenna

labial palpus

proboscis

Metamorphosis complete, pronounced; larvae (caterpillars) mandibulate with paired false legs on some abdominal segments in addition to normal thoracic legs; pupae usually enclosed in a cocoon or earthen cell.

BEETLES
Order Coleoptera

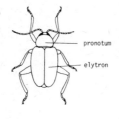
pronotum
elytron

Minute to large; hard-bodied; aquatic or terrestrial. Head usually prominent; with well-developed mandibulate mouthparts; antennae variously modified, often with apical segments enlarged; ocelli nearly always absent. Prothorax distinct from rest of thorax; 2 pairs of wings, the front pair (elytra) thickly sclerotized, almost always meeting in a straight line down the middle of the back, hindwings membranous, usually folded beneath the elytra when at rest. Metamorphosis complete; larvae mandibulate, variously shaped; pupae usually in cells in ground or plant material.

STYLOPS
Order Strepsiptera

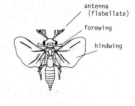
antenna
(flabellate)
forewing
hindwing

Small; larvae and females endoparasitic in other insects. Males free-living; eyes well developed; antennae with one or several segments prolonged laterally (flabellate); mouthparts reduced; prothorax and mesothorax small, metathorax large; forewings small,

club-shaped; hindwings large.
Female larviform, remaining in
puparium; with antennae and
mouthparts absent; head and thorax
fused; abdomen a large sac with 3-5
sexual openings giving birth to larvae (viviparous). Metamorphosis
complete, complex; first stage larvae active, long-legged; later stages
grublike; pupae in host.

ANTS, WASPS, BEES
Order Hymenoptera

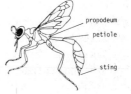

Moderate-sized, small or minute,
rarely large, with varied plant-
feeding (phytophagous), predatory,
or parasitic habits. Four membranous wings, the fore pair larger and
more complexly veined than hind;
mouthparts mandibulate but usually
modified for lapping; abdomen
usually with six or seven visible
segments, the first true segment
(propodeum) fused with the thorax,
not forming part of apparent abdomen; ovipositor often elongate,
sawlike or drill-like, or modified
into a sting. Metamorphosis complete; larvae caterpillarlike or grublike; pupae free, sometimes enclosed
in a cocoon.

SYSTEMATIC TREATMENT

PRIMITIVE WINGLESS INSECTS
Apterygota

Members of the Apterygota lack wings in all stages. This feature is considered to be primitive, these forms representing relicts of the most ancient insects descended from centipede-like ancestors. Vestiges of abdominal legs also evidence this ancestry and separate the Apterygota from all the winged orders. (Although fleas, lice, ants, and a few other higher insects are wingless, the condition is clearly a secondary loss related to parasitic habits or to other ecological specializations.)

All Apterygota grow without metamorphosis; the immatures resemble the adults except for smaller size and undeveloped reproductive organs.

PROTURANS
Order Protura

1. Proturans are primitive wingless insects, virtually unknown to the layman because of their extremely small size (BL 0.5–2 mm) and secretive habits. They lack eyes and antennae, the function of the latter being assumed by the front pair of legs which are held forward in a probing position. The basal 3 abdominal segments bear ventral, segmented appendages called styli. Although seldom seen even by entomologists, enormous populations may inhabit areas of moist soil, leaf mold, rotting wood, and similar situations. Only about 10 species have been found in California, the commonest probably being the **(1) California Telson-tail** *(Nipponentomon californicum)*, found throughout the state in the upper soil layers.

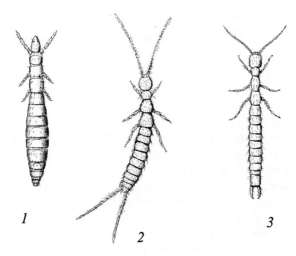

Figs. 1–3 Proturan: 1, California Telson-Tail, *Nipponentomon californicum*. Diplurans: 2, Folsom's Campodeid, *Campodea folsomi;* 3, Large Japyx, *Japyx diversiunguis.*

Reference

Tuxen, S. L. 1964. *The Protura*. Paris: Hermann Co.; 360 pp. A revision of the species of the world with keys for determination.

DIPLURANS
Order Diplura

2, 3. Diplurans are wingless insects restricted to moist habitats. Recognition characters are small size (BL 6 mm or less), pale-colored body, compound eyes absent, tarsi 1-segmented, and pair of conspicuous appendages (threadlike or pincherlike) from the tip of the abdomen. Diplurans live primarily in soil and leaf mold but may also occur under bark or objects lying on the ground. Several species are common in California, such as **(2) Folsom's Campodeid** *(Campodea folsomi)* and **(3)** the **Large Japyx** *(Japyx diversiunguis)*. One rare family (Anajapygidae; posterior appendages segmented but very short) has its sole North American representative *(Anajapyx hermosus)* in Placer County.

SPRINGTAILS
Order Collembola

Springtails, like proturans and diplurans, are primarily members of the soil community, but they are of more importance because of their much greater diversity and numbers. Most springtails are readily recognized by a forked, tail-like appendage (furcula) which arises towards the rear of the abdomen and which the insect snaps against the substratum, springing itself into the air. A small, tubular lobe (collophore) projects between the legs from the underside of the first abdominal segment. Other characteristics are short abdomen and antennae (6 or fewer segments), small size (BL normally less than 6 mm), and absence of cerci.

Habitats are soil, leaf litter, decaying wood, in fungi, near and on water, and similar damp places. A few species, called "snow fleas" (*Onychiurus cocklei*, *Achorutes nivicola* in California), swarm in enormous numbers on snow. The following are only a few of about 150 species in the state.

Reference

Christiansen, K. 1964. Bionomics of Collembola. *Ann. Rev. Ent.* 9:147–178.

4. Varied Springtail. *Isotoma viridis*. ADULT: Maximum BL 6 mm; extremely varied color pattern, base color almost any shade, commonly green; body form elongate; antennae long and furcula present; longest body hairs strongly barbed or toothed. Common in damp places. RANGE: Widespread through North America.

5. Culture Louse. *Onychiurus folsomi*. ADULT: BL 1.8–2.1 mm; body cylindrical; eyeless; antennae short; cuticle with numerous minute spots (pseudocelli) through which a protective fluid is released. Abundant in earthworm, fruit fly, and other cultures as well as moist places in nature, such as under cattle dung. RANGE: Of general occurrence.

6. Laguna Marine Springtail. *Entomobrya laguna*. ADULT: BL 6–7 mm; usually with bluish mottling on gray ground color; furcula well developed; numerous spatulate hairs on head and elongate body. Lives in rock hollows and crevices

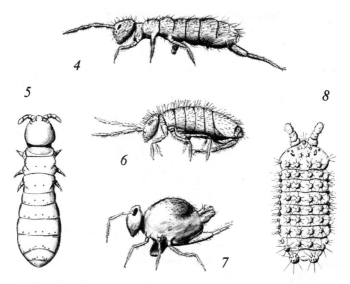

Figs. 4–8 Collembolans: 4, Varied Springtail, *Isotoma viridis;* 5, Culture Louse, *Onychiurus folsomi;* 6, Laguna Marine Springtail, *Entomobrya laguna;* 7, Lawn Springtail, *Bourletiella arvalis;* 8, Obese Springtail, *Morulina multatuberculata.*

in the marine intertidal zone where it survives complete submergence. RANGE: Rocky portions of entire coast. A structurally similar but yellowish terrestrial species, *E. unostrigata,* is common in lawns.

7. Lawn Springtail. *Bourletiella arvalis.* ADULT: BL 1.5 mm; compact, globular body with posterior lobes above furcula; segmentation indistinct; generally light yellow, without markings. Male and female dissimilar. Male with enlarged bristles at end of abdomen, lacking in female. Common in lawns. RANGE: Widespread.

8. Obese Springtail. *Morulina multatuberculata.* ADULT: BL up to 5 mm, bulky; bluish white, with conspicuous bristle-bearing tubercles over entire body; furculum absent; very short, 4-segmented antennae. Common in spring under stones, logs, and other objects lying on leaf mold and in other moist soil. RANGE: Coastal and in northern counties; found primarily in wooded, hilly areas.

SILVERFISH AND BRISTLETAILS
Order Thysanura

Two major groups make up the order Thysanura: the silverfish (so-called because of their shiny gray color and slippery, scaly surface) and the bristletails or rockhoppers (the name referring to their habit of jumping).

Silverfish are better known from their habit of infesting homes and buildings where they feed on a wide variety of dry organic materials such as paper, bookbinding paste, starch in clothing, fabrics, dried meat, cereals, etc. Household species probably originated in warmer countries and survive in California primarily in heated buildings. There are also many native species in such habitats as sand dunes. Anatomical characteristics of this suborder are the small, widely separated eyes, anteriorly projecting mouthparts, somewhat flattened body, and middle and hind coxae with stylelike projections. Locomotion is primarily by crawling.

Bristletails, or rockhoppers, are usually found out of doors: under stones, beneath bark, in leaf litter, etc. They feed on lichens, terrestrial algae, molds, decaying fruits, and dead insects. Distinguishing structures are the large, contiguous eyes, ventrally projecting mouthparts, rounded body (in cross-section), and middle and hind coxae without stylelike projections.

Species of both groups also possess eversible, bladderlike vesicles on the underside of the abdomen which are used to sponge up moisture from the substratum. Most also are secretive and shun the light (except *Machilinus* which is diurnal). A body covering of scales is another general feature of the order, except in *Tricholepidion*.

Reference

Wygodzinsky, P. 1972. A revision of the silverfish (Lepismatidae, Thysanura) of the United States and the Caribbean area. *Amer. Mus. Novitates,* No. 2481; 26 pp.

Family Lepidothrichidae

9. Venerable Silverfish. *Tricholepidion gertschi.* ADULT: BL 12 mm; reddish brown or yellowish; lacks scales. Lives in cool crevices and under decaying bark and in rotten logs of

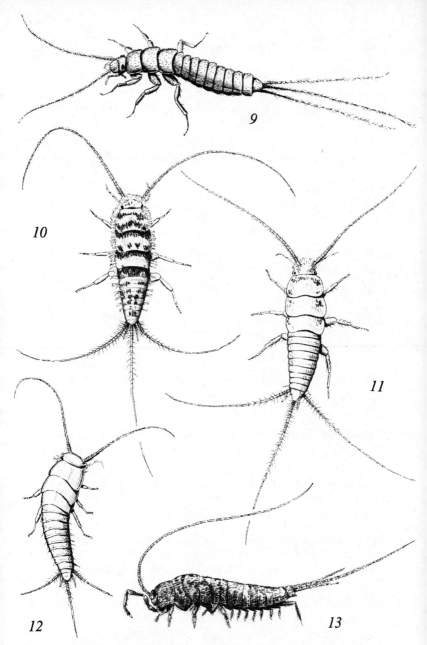

Figs. 9–13 Thysanurans: 9, Venerable Silverfish, *Tricholepidion gertschi;* 10, Firebrat, *Thermobia domestica;* 11, Long-Tailed Silverfish, *Ctenolepisma longicaudata;* 12, Common Silverfish, *Lepisma saccharina;* 13, Rockhopper, *Mesomachilis* species.

Redwood and Douglas Fir. RANGE: Restricted to the Redwood-mixed conifer forest of the Coast Range of Mendocino County; known only from localities along the Eel River. This is a living fossil. Anatomically it exhibits many primitive characteristics compared to other silverfish, and it closely resembles the probable hypothetical ancestor of higher insects. The family was first described on the basis of a similar, extinct species found fossilized in Baltic amber, 30,000,000 years old.

Family Lepismatidae

10. Firebrat. *Thermobia domestica.* ADULT: BL to 13 mm; color of body two-toned, scales brown and pale giving a mottled pattern; maxillary palpus with 6 segments; upper body hairs in tufts of combs. This species is notable for its preference for hot surroundings and is commonly found around ovens, bakeries, heating pipes, furnaces, boilers, etc., where temperatures often range from 32–40°C. RANGE: Throughout warmer parts of California. The **Sand Dune Silverfish** *(Leucolepisma arenaria)* is similar in appearance, being pale tan, but without strong markings on the back. It is a native species, common in the deserts of California.

11. Long-Tailed Silverfish. *Ctenolepisma longicaudata.* ADULT: BL up to 15 mm; uniform slate gray with outer bristles of abdomen arranged in 3 rows of combs on each side; tail about as long as body. Larger than, but with food and other habits similar to the Common Silverfish (see **12**). RANGE: Widespread in urban and industrial situations.

12. Common Silverfish. *Lepisma saccharina.* ADULT: Maximum BL 12 mm. Uniform gray with a silvery sheen; like the Long-tailed Silverfish but body hairs single or in small groups; maxillary palpus with 5 segments; tails shorter than length of abdomen. Primarily a house-infesting insect, often doing great damage to paper and other stored organic products. Prefers moister and cooler places than the (**10**) Firebrat (21–27°C optimal). RANGE: Cosmopolitan and widespread in California, in domestic and commercial buildings. California's most common native species is the **Spined Silverfish** *(Allac-*

rotelsa spinulata), a dark to pale brown species having long, spinelike body hairs and moderately long tails, all 3 of about equal length (BL to 12 mm). Common in leaf litter, rotting logs, and in sandy areas along the coast, in the foothills of the Coast Ranges and Sierra Nevada, and in the Central Valley.

Family Machilidae

13. Rockhoppers. *Mesomachilis, Pedetontus.* ADULTS: The several species in these and related genera of so-called "jumping silverfish" are separable only by the expert. (See introduction to this order for the features that distinguish them from true silverfish.) Occasional specimens enter homes, but rockhoppers normally are seen only outdoors. RANGE: General throughout the state; most common in dry places, such as rocky or sandy habitats, and in leaf litter.

WINGED INSECTS
Pterygota

The vast majority of insects are members of Pterygota and typically possess wings in the adult stage. Abdominal appendages other than the genitalia or terminal cerci are never present in the adults. All species mature through a series of body changes (metamorphosis) in which the immatures may exploit different habitats and differ profoundly in body form from the adults.

MAYFLIES
Order Ephemeroptera

Adult mayflies are small to medium-sized insects with glassy-appearing wings, with frail bodies and 2 or 3 long, threadlike tails. They are commonly seen in swarms, engaged in peculiar, dipping flights above ponds and streams where they are easy prey for dragonflies and fish. The ordinal name derives from the ephemeral life of adults which lack functional mouthparts and do not feed. Each lives but a few hours or a day or two, an ignoble end for the immature which has taken months to mature.

Mayflies first leave the water as pre-adult winged forms (subimagos), recognized by their milky wings and poor flight abilities. After about a day the subimagos molt to become the clear-winged, graceful, flying adult. These are the only insects that undergo molting in a winged state.

The aquatic immatures, called naiads, are remarkably diverse in form. They breathe by means of articulated, flaplike gills located at the sides of the abdominal segments. Because most are herbivorous and abundant in various freshwater habitats, they occupy a vital role near the base of the food chain. They constitute a staple in the diet of many predators, especially fish.

For centuries anglers have recognized the importance of mayflies as fish food and have used the emergence periods ("hatches") of the subimagos ("duns") and adults ("spinners") to time fishing efforts as well as to employ the insects or feathered replicas of them as bait ("flies"). The 3 kinds mentioned are common representatives of the approximately 170 species in 7 families which occur in the state.

Reference

Edmunds, G. F. Jr., S. L. Jensen, and L. Berner. 1976. *The Mayflies of North and Central America.* Minneapolis: University of Minnesota Press; 330 pp.

Family Ephemerellidae

These are California's most common mayflies, with more than 80 species recorded in the state.

14, 15. Blue-winged duns. *Ephemerella.* There are about 24 species in this genus in California. ADULTS (**14**): Variable in size (BL 5–15 mm), color, and markings; most species with brownish body and clear wings; males with divided eyes, lower portion small and dark, upper portion large and cone-shaped. NAIADS (**15**): BL 5–20 mm; most with enlarged, spined femora; legs and gills with hair fringes or thoracic and abdominal plates armed with tubercles and spurs. Most species are found under rocks in moderate currents; others on silt bottoms or among trash and debris. RANGE: Rivers and streams throughout most of the state.

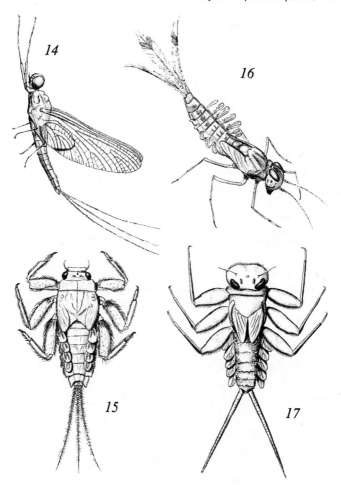

Figs. 14–17 Mayflies: 14, Blue-Winged Dun, *Ephemerella* species (adult); 15, *Ephemerella* species (naiad); 16, Pacific Spotted Mayfly, *Callibaetis pacificus* (naiad); 17, Flat Nymph Mayfly, *Epeorus* species (naiad).

Family Baetidae

Members of the Family Baetidae are among the most prolific of the pond herbivores and their numbers may reach prodigious proportions. They serve as the major sustenance for associated carnivores, both vertebrate and invertebrate. Although species of the genus *Baetis* are the most commonly encountered mayflies, the following species is a characteristic representative of this family in California.

16; Plate 1a. Pacific Spotted Mayfly. *Callibaetis pacificus.* ADULT: BL 9–10 mm. Male (**Plate 1a**) has dark brown pattern on abdominal segments 5, 6, and 9, with large pale areas; forewing with a wide, brown anterior margin. Female has tan, dark markings on back of abdominal segments 2, 3, 5, 6, and 9; forewing entirely clear. NAIAD (**16**): A still-water form and an excellent swimmer; body frail, elongate (BL 6–10 mm). RANGE: Southern California and the Coast Ranges north to Marin County.

Family Heptageniidae

17. Flat nymph mayflies. *Epeorus, Heptagenia,* and *Rhithrogena.* ADULTS: BL 6–11 mm, WL 10–16 mm; wings entirely clear; body light brown with paler legs. NAIADS: Many species in these genera have a flattened body and a rounded, depressed head with the eyes on top; gills broad and flaplike, the first and last pairs often meeting beneath the body and overlapping the lateral ones to form the rim of a "suction cup" (abdomen acting as the piston) which assists the insect in its hold upon smooth rocks in swift water. Long fringes of hairs on the legs enhance the effect. RANGE: Most California species live in the Sierra Nevada, in high-elevation streams where naiads cling to rocks in fast currents.

DRAGONFLIES AND DAMSELFLIES
Order Odonata

Dragonflies and damselflies are familiar insects found around ponds and streams where the immatures, called nymphs or naiads, develop in the water. Both adults and naiads are predaceous: adults catch small flying insects, such as mosquitoes and other gnats, while the naiads eat a wide variety of aquatic animals, even tadpoles and small fish. The immature forms are usually dull colored or hairy to catch mud and debris and to match the surroundings at the bottom of their habitat. From their hiding places, they snatch up their prey by quick extensions of a long, jointed lower lip (or labium). About 100 species are known in California.

References

Corbet, P. S. 1963. *A Biology of Dragonflies.* Chicago: Quadrangle Books; 247 pp.

Needham, J. G., and M. J. Westfall. 1955. *The Dragonflies of North America.* Berkeley and Los Angeles: University of California Press; 615 pp.

DAMSELFLIES
Suborder Zygoptera

The damselflies are frail-bodied compared to dragonflies and have the eyes protruding at the sides on short stalks. They are weak fliers, flitting about and perching on vegetation in the immediate area of water. The wings are held together when at rest, butterfly-style above the abdomen. Eggs are inserted into tissues of rushes or other vegetation, usually just below the surface of the water. The nymphs breathe by leaf-shaped, external gills at the posterior end. About 40 species are recorded in the state.

Family Calopterygidae

18. Common Ruby-spot. *Hetaerina americana.* This is one of California's largest and most beautiful damselflies, unmistakable with its bright red, basal wing spots. ADULT (**18**): BL 37–50 mm; body bronzy colored in both sexes; wing spots in male brilliant carmine red, diffuse and dull reddish, amber, or brownish in female. NAIAD: Protectively colored, mottled brown; clinging to debris, plants, or rocks at the edge of slow currents. RANGE: Most often seen along slow streams or around springs, primarily in dry parts of the state, ranging north in the Central Valley to Butte County.

Family Coenagrionidae

19. Bluets. *Enallagma.* These are the common bright blue and black damselflies that are seen around all kinds of aquatic habitats where there is shallow, fresh water with abundant vegetation. ADULTS: Medium-sized (BL 28–40 mm); males predominantly bright blue; females vary, blue or gray or brown—with black markings and with colorless wings. They fly immediately above the surface of the water, often perching

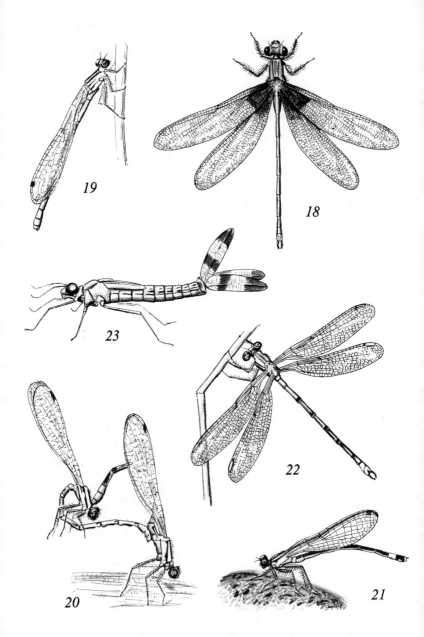

Figs. 18–23 Damselflies: 18, Common Ruby-Spot, *Hetaerina americana* (adult); 19, Bluet, *Enallagma* species; 20, Vivid Dancer, *Argia vivida* (mating pair); 21, Fork-Tail, *Ischnura* species; 22, California Spread-Wing, *Archilestes californica* (adult); 23, *A. californica* (naiad).

on emergent vegetation, rarely wandering far from the shore. NAIADS: Long, pale green or brown, with gills tapering to a point. They actively clamber among submerged plants in lakes and ponds. RANGE: Five species are common and widespread in California, all quite similar in appearance and all more or less widespread, although usually not all occur together.

20; Plate 1b. Dancers. *Argia*. ADULTS: Moderately large (BL 28–50 mm); narrow-winged, blue, violet, or tan with black markings resembling those of the bluets *(Enallagma)* but with larger bodies and much longer leg spines. Dancers are fast, jumpy damselflies, which often rest on bare ground or rocks in contrast to bluets which prefer to perch on vegetation. NAIADS: Short, thick-skinned, often dark brown. Slow-moving among submerged vegetation or rocky recesses near the edges of ponds and streams. Posterior caudal gills short, arched-shaped, often with a pale transverse band. RANGE: Nine species occur widely in California, the most common being the **(20, Plate 1b) Vivid Dancer** *(A. vivida),* a ubiquitous species.

21. Fork-tails. *Ischnura*. These damselflies also occur near still, fresh water, often at margins of ponds rather than along running streams with the bluets. The fork-tails are so-called because the males have a bifurcate process on the last dorsal plate of the abdomen. ADULTS: Small (BL 22–32 mm) with bright green or blue spots. In contrast to *Enallagma* and other damselflies, the abdomen is mostly black with a conspicuous color band preceding the tip. Females in some species have two color forms, one resembling the male and the other less brightly colored. NAIADS: Long, pale green, resembling *Enallagma* from which they differ by having 6 instead of 5 antennal segments. RANGE: Six species of *Ischnura* occur widely in California, 3 common, and all similar in appearance.

Family Lestidae

22, 23. California Spread-wing. *Archilestes californica*. Lestids differ from other damselflies in that they usually spread out the wings as dragonflies do when perched. ADULT (**22**): Large (BL 48–55 mm); body black and tan, wings colorless.

The eggs are inserted in regular groups of 6 into stems of woody plants, such as willow, just under the bark and sometimes several meters above the water surface. NAIAD (**23**): Slender, with elongate, banded anal gills; climbing on vegetation in still water, such as pools at the margins of streams. A related species, the **Giant Spread-wing** (*A. grandis*) is the largest damselfly in California, BL reaching 60 mm.

DRAGONFLIES
Suborder Anisoptera

Dragonflies are among the most conspicuous and familiar of all insects, owing to their large size, bright colors, and strong, swift flight. They have long, slender bodies and enormous compound eyes that enable perception of prey, which is captured on the wing. Males of many species course around their habitats in a territorial fashion, engaging in combat with other insects in their size range, especially males of their own species. Their aerial maneuvering is extraordinarily impressive to every naturalist, particularly their uncanny ability to alter their position with effortless grace, even disdain, so as to remain just out of reach of the collector's net.

The naiads are varied in form and in their aquatic habitats. Although they appear awkward when climbing over debris or burrowing in bottom mud, by forcibly squirting water from the rectum, they can suddenly jet forward to escape enemies or to seize their victims. Oxygen is absorbed from the water by internal rectal gills. Development requires a few months to four years. Egg placement by females varies, from unselective scattering on the water to insertion into mud below shallow water or into plant tissues.

All 7 North American families occur in California, represented by 64 species.

SKIMMERS
Family Libellulidae

24, 25, 26. White tail skimmers. *Libellula*. These are showy species with a powdery white abdomen, black spotted wings, and additional white spots in the male. The naiads are hairy, sluggish, sprawling in bottom silt and debris. The (**24**) **White Tail** (*L. lydia*) occurs throughout California at low to moderate

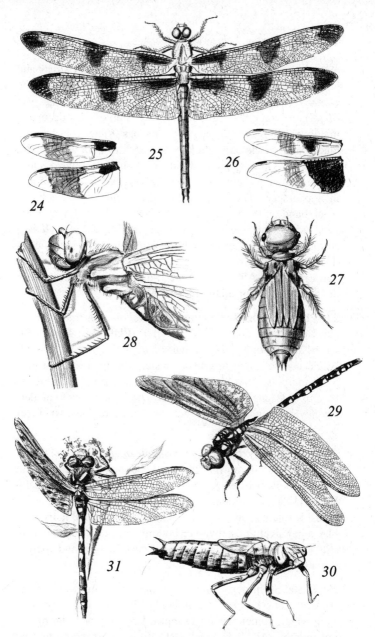

Figs. 24–31 Dragonflies: 24, White Tail, *Libellula lydia* (left-side wings); 25, Ten Spot, *L. pulchella;* 26, Widow, *L. luctuosa;* 27, Big Red Skimmer, *L. saturata* (naiad); 28, Pastel Skimmer, *Sympetrum corruptum* (head and thorax); 29, Common Blue Darner, *Aeshna multicolor* (adult); 30, *A. multicolor* (naiad); 31, Yellow-Backed Biddie, *Cordulegaster dorsalis.*

elevations. The male has a short, longitudinal black streak and white spot at the base and a broad dark band beyond the middle of each wing; BL 49–51 mm. The (**25**) **Ten Spot** *(L. pulchella),* a larger (BL 52–57 mm) but rarer species occurring mainly in the Sierra Nevada and the Coast Ranges, has partial black bands at the base, middle, and apex of both wings, with intervening white spots on the male, 2 on the forewing and 3 on the hindwing. Females of these two species are quite similar in appearance. The (**26**) **Widow** *(L. luctuosa)* has a broad black band on the basal half of the wing, followed by a white band across the middle in the male; BL 42–50 mm. It is a common pond species through much of California but is missing east of the Sierra Nevada and in the Coast Range south of San Francisco. All 3 species are usually found hovering over ponds, marshes, or other sluggish water.

27; Plate 1c. Big Red Skimmer. *Libellula saturata.* One of our commonest dragonflies. ADULT (**Plate 1c**): Large (BL 52–61 mm), body and basal half of wings bright rust colored. Adults course slowly up and down sluggish streams and around pond margins. NAIAD (**27**): Hairy; squats amid ooze at the bottom of stagnant ponds. RANGE: Throughout low to moderate elevations in California, including desert regions, wherever water accumulates.

Plate 1d. Dusty Skimmer. *Sympetrum illotum.* ADULT: Small (BL 38–40 mm); brilliant red abdomen contrasting with the dull brownish red of the wings. Dusty Skimmers usually are seen coursing around standing water, pond margins, along roadside ditches, and so forth. NAIAD: Stocky and hairy and, like other libellulids, lives in bottom trash and vegetation in still water. RANGE: Throughout Coast Ranges and Sierran foothills at low elevations.

28. Pastel Skimmer. *Sympetrum corruptum.* In contrast to its relatives, this species is commonly observed long distances from water, sometimes migrating in large numbers during the fall months. ADULT: Small (BL 39–42 mm); body with an intricate color pattern, pale orange brown with pale bluish spots along the sides; wings colorless except for orange-tinged veins. NAIAD: Compact, relatively slender; lives amidst vegetation and bottom trash in still and very slow-moving waters of ponds and minor streams. RANGE: Widespread in California, occurring at all elevations in arid and moist regions.

DARNERS
Family Aeshnidae

Plate 1e. Common Green Darner. *Anax junius.* ADULT: Large (BL 68–80 mm); strong-flying species with bright green thorax, blue abdomen, and ochreous-tinged wings. NAIAD: Elongate, smooth, decorated with patterns of green and brown; lives among submerged pond vegetation. RANGE: One of the most widespread North American dragonflies and common throughout the state. A related species, **Walsingham's Darner** *(A. walsinghami)* occurs in the southern part of the state; it is the largest North American dragonfly (male BL attaining 100–115 mm); wings do not become yellow with age in this species.

29, 30. Common Blue Darner. *Aeshna multicolor.* ADULT (**29**): Moderately large (BL 68–72 mm); body predominately blue; wings colorless. Females deposit eggs into stems of aquatic plants. NAIAD (**30**): Streamlined, neatly patterned in longitudinal streaks of green and brown; lives in submerged vegetation of ponds, marshes, sluggish streams, and similar habitats, near which the adults are most often seen. RANGE: Perhaps the commonest dragonfly in lowland areas throughout California. A related species, *A. palmata,* is common in montane areas.

BIDDIES
Family Cordulegastridae

31. Yellow-backed Biddie. *Cordulegaster dorsalis.* ADULT: Among the largest of California's dragonflies (BL 70–85 mm); body banded black and yellow; wings colorless. NAIAD: Body stout, rough, hairy, and often covered with green algae; lives in mud at the bottom of flowing streams where it burrows by kicking silt with the hind legs and descends by squirming with the whole body. RANGE: Often found coursing along woodland canyons in the Coast Ranges and Sierran foothills. A desert form lives around springs in Inyo County.

STONEFLIES
Order Plecoptera

Stoneflies are aquatic insects whose nymphs require moving water for development. Accordingly, the weak-flying adults are usually found near mountain streams. They are dull colored with 4 membranous wings that fold over the back when at rest; most species are gray or brown; others are yellowish or green. They are commonly found on vegetation or resting on rocks or tree trunks. Although somewhat resembling winged termites, alderflies, and dobsonflies, stoneflies are distinguished by having fan-shaped hindwings with many veins and 3-segmented tarsi.

The nymphs have two long cerci or tails and often are provided with fringes of long, silky hairs on the wing pads, legs, or tails. They obtain oxygen from the water with hairlike or branched gills on the underside of the thorax or abdomen.

There are about 100 species in California, with all 6 North American families represented. They vary in size from 4 to 50 mm in body length, up to 60 mm to the tips of the wings.

References

Jewett, S. G. 1960. The stoneflies (Plecoptera) of California. *Calif. Insect Survey Bull.* 6(6): 125–177.

Needham, J. G., and P. W. Claassen. 1925. *A Monograph of the Plecoptera or Stoneflies of America North of Mexico.* Lafayette, Indiana: Thos. Say Foundation; 397 pp.

Family Perlidae

32; Plate 1f. Yellow-banded Stonefly. *Calineuria californica.* ADULT (**Plate 1f**): BL 23–31 mm; wings and body brown, body with yellow markings, including the conspicuous median line on top of the thorax. It flies from April to July. NYMPH (**32**): Thorax with profusely branched gills at lower angles, anal gills lacking; eyes located near hind margin of head; a yellowish triangle in front of eyes. The nymph of a similar species, *C. pacifica,* has anal gills. RANGE: *C. californica* and *C. pacifica* are both widespread, from Los Angeles County northward.

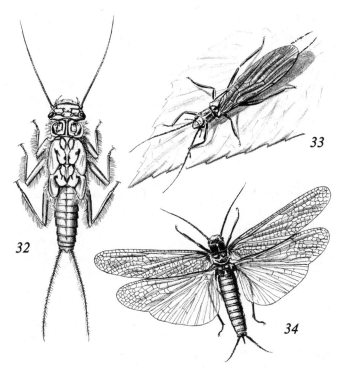

Figs. 32–34 Stoneflies: 32, Yellow-Banded Stonefly, *Calineuria californica* (nymph); 33, Early Brown Stonefly, *Capnia* species; 34, California Giant Stonefly, *Pteronarcys californica*.

Family Capniidae

Adults of many capniids are active very early in the year and are often referred to as "winter stoneflies." They are commonly observed around snowbound streams, alighting on rocks, tree trunks, and even in large numbers on the snow itself.

33. Early brown stoneflies. *Capnia.* Members of this genus are among the commonest stoneflies in California; seventeen species are represented, all similar in appearance. ADULTS: Small (BL 6–10 mm); blackish, with clear, smoky or brown wings and long cerci or tails; observed in early spring, mostly between February and April in California, with a few species as late as June. NYMPHS: Inner margin of hindwing pad with a median notch; the nymphs of most species have not been associated with the adults or described. RANGE: Several species are widespread in the Sierra Nevada and the Coast Ranges. A recently discovered, unique species in this genus is the **Tahoe Wingless Stonefly** *(C. lacustra);* adults are without wings and live their entire lives at depths in excess of 60 m (195 ft) on the bottom of Lake Tahoe.

Family Pteronarcyidae

34. California Giant Stonefly. *Pteronarcys californica.* This is one of the largest stoneflies in North America. It can be found around shaded streams between April and July. ADULT: Male (BL to wing tip 35–45 mm); female larger (BL to wing tips up to 60 mm); gray or brown with a reddish median line on the thorax and pale or reddish spots at the sides. NYMPH: Abdominal segments 1–2 with branched gills on the underside; thorax with long lateral processes and pointed wing pads. RANGE: Sierra Nevada and the Coast Ranges from Kern and Santa Clara counties northward. A similar species, *P. princeps,* is slightly smaller and occurs in the same area as well as the mountains of southern California.

COCKROACHES
Order Blattodea

The domestic cockroaches are all too familiar insects. Fortunately they are only a few species of a large and interesting, but mainly tropical, group. Cockroaches are easily recognized by a compressed body, large, oval, shieldlike prothorax almost hiding the head, very long, slender antennae, aversion to light, a glossy, waxy surface, and catholic food habits. All form a purse-shaped egg capsule which is often seen protruding from the female's abdomen, where it may be carried up to a month.

Although cockroaches are widely considered as dirty, they are probably of little importance in disease transmission except when very abundant under filthy conditions.

The first 6 of the following species are cosmopolitan household pests long associated with man; all are considered to be African in origin, probably having spread to America in the slave-ships either directly *(Periplaneta* and *Supella)* or via Europe *(Blattella* and *Blatta)*. The others characterized below are representative of our meager native fauna of 5 or 6 species.

Reference

Rehn, J.W.H. 1951. Classification of Blattaria as indicated by their wings (Orthoptera). *Mem. Amer. Ent. Soc.* 14; 134 pp.

35, 36. German Cockroach or **Croton Bug.** *Blattella germanica.* ADULT (**35**): Smallish (BL 16 mm); with full set of wings extending beyond the tip of the abdomen in both sexes; light brown with parallel, dark, longitudinal stripes on head shield. Gregarious and often develops enormous populations; prefers bathrooms, kitchens, and commercial establishments. EGG CASE (**36**): Length 8 mm; light brown, soft; elongate, rectangular; 30–40 egg chambers, conspicuously marked externally. RANGE: Widespread throughout coastal areas, Central Valley, and Sierran foothills; may occur in any domestic environment. The related **Field Cockroach** *(B. vaga)* is normally an outdoor species in the Central Valley and deserts but can become a pest indoors.

37, 38. Oriental Cockroach. *Blatta orientalis.* ADULT (**37**): BL to 32 mm; male with brownish wings which do not quite reach end of abdomen; female wingless; body shining black. Because it is often attracted to moist areas (bathroom toilet bowls, bathtubs, water pipes, kitchen sinks, etc.), this species has acquired the name "water bug"; gregarious, but forms only small colonies. EGG CASE (**38**): Largest of the domestic species (length 10 mm); black, hard, and smooth; purselike (rounded ends); usually with about 16 egg chambers. RANGE: Statewide in domestic habitats; survives well outdoors in warm weather.

39, 40. Brown-banded Cockroach. *Supella supellectilium.* ADULT (**39**): Smallest domestic form (BL no more than 14 mm); generally brown with two pale transverse bands crossing the wings; wings fully developed in both sexes, but not quite extending to tip of abdomen in female. Solitary and the least common of the domiciliary roaches. EGG CASE (**40**):

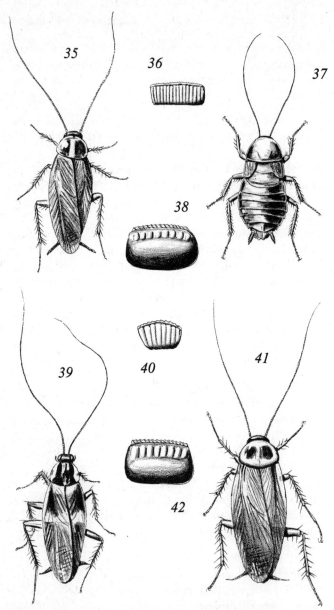

Figs. 35–42 Cockroaches: 35, German Cockroach, *Blattella germanica* (adult); 36, *B. germanica* (egg case); 37, Oriental Cockroach, *Blatta orientalis* (adult female); 38, *B. orientalis* (egg case); 39, Brown-Banded Cockroach, *Supella supellectilium* (adult); 40, *S. supellectilium* (egg case); 41, American Cockroach, *Periplaneta americana* (adult); 42, *P. americana* (egg case).

Smallest of indoor species (length 4 mm); yellowish to reddish brown; rectangular (lower ridge convex); 14 egg chambers, well-marked externally; usually stuck to vertical surfaces in clusters. RANGE: Occurs only in warmer parts of our state.

41, 42. American Cockroach. *Periplaneta americana.* ADULT (**41**): Large—but not as immense as barroom tales—maximum BL 50 mm; both sexes fully winged; reddish to dark brown head shield, with pale yellowish periphery and dark brown splotches in the center. EGG CASE (**42**): Length 8 mm; black, hard, and smooth; purselike (rounded ends); a series of teeth along upper ridge of case; 14–16 egg chambers. RANGE: Cosmopolitan and statewide. The most prevalent house roach in California. Usually occurs in small numbers. The **Australian Cockroach** *(P. australasiae)* occurs sporadically in California in urban situations; it is recognized by the bright yellow border on the head shield and elongate, yellow shoulder triangles.

Plate 1g. Sand roaches. *Arenivaga, Eremoblatta.* ADULTS: Males fully winged, the wings mottled pale brown or translucent; BL 20–30 mm to wing tips; commonly attracted to lights at night in the desert or other sandy areas. Females wingless, oval, and strongly convex (form suggestive of prehistoric Trilobites); burrow in sandy soil and dunes; presence evidenced by serpentine surface ridges. RANGE: Several species in the state, limited to deserts, Central Valley, and southern coastal dunes. (**Plate 1g**) **Hairy Desert Cockroach** *(Eremoblatta subdiaphana)* is distinguished by the translucent wings of the male and the yellowish hairs of the body in the female from *Arenivaga,* which has opaque, leathery, and often mottled wings and relatively hairless females. The only other common native roach is the **Western Wood Cockroach** *(Parcoblatta americana).* ADULT: Shining, dark brown, the wingless female much darker than male; wings well developed and extending beyond the abdomen in the males, which are often attracted to lights. Found in old logs and under rocks in chaparral areas. RANGE: Foothills to middle elevations throughout the state.

TERMITES
Order Isoptera

Termites are well known because of their destructive feeding on structural wood. They are social insects, with a well-defined caste system, and many species develop immense colonies. Although termites are most abundant, diverse, and destructive in tropical regions, they sometimes cause serious damage to structural timbers in temperate areas, stimulating the growth of a considerable industry associated with control and preventive measures.

Termites are small and soft-bodied, normally pale or white (sometimes called "white ants"), with biting mouthparts and incomplete metamorphosis. Some forms are blind, and many species develop seasonal, winged sexual forms. The 4 similar wings are long and narrow; they fold flat over the back when at rest and often are lost following the mating flight.

The colonies live either in wood above ground or in the ground, feeding on buried wood, decaying roots, and so forth, or on wood above ground which is made accessible by means of mud tunnels. In the latter situation, moisture is required, and these subterranean termites are usually the destructive species in California. There are about 15 species known in the state, of which about 10 are desert species of little or no significance to man.

References

Ebeling, W., and R. Pence. 1965. *Termite Control*. Calif. Agric. Exp. Sta., Extension Service, Circ. 469 (revised). Berkeley: University of California Press.

Krishna, K., and F. Weesner, eds. 1969–1970. *Biology of Termites*. New York: Academic Press; 2 vols., 1200 pp.

Weesner, F. M. 1965. *The termites of the United States. A Handbook*. Elizabeth, N.J.: National Pest Control Assoc.; 70 pp.

Family Hodotermitidae

43, 44, 45. Pacific Dampwood Termite. *Zootermopsis angusticollis*. This is the largest yet one of the least destructive of California's termites. ADULT: Colonies consist of large (BL 16–24 mm), yellowish soldiers with red-brown heads and conspicuous black mandibles (**45**); smaller reproductives which

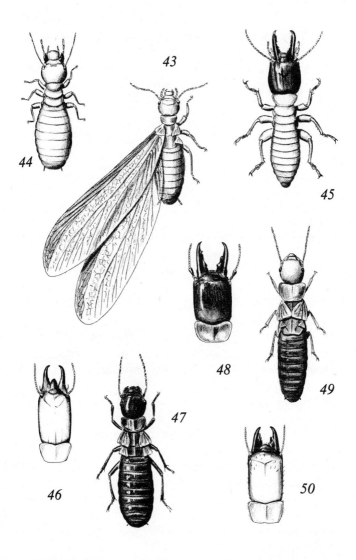

Figs. 43–50 Termites: 43, Pacific Dampwood Termite, *Zootermopsis angusticollis* (reproductive); 44, *Z. angusticollis* (immature reproductive); 45, *Z. angusticollis* (soldier); 46, Western Subterranean Termite, *Reticulitermes hesperus* (soldier head); 47, *R. hesperus* (dealated reproductive); 48, Western Drywood Termite, *Kalotermes minor* (soldier head); 49, *K. minor* (dealated reproductive); 50, Desert Dampwood Termite, *Paraneotermes simplicicornis* (soldier head).

may be fully winged (**43**), short-winged, or apterous; winged form light brown with a short body (BL 9–10 mm) and long (20–24 mm), dark brown, heavily veined wings. The wings are shed following the mating swarms and establishment of new colonies in August and September. There are no true workers, this function being carried out by immature forms of the reproductives (**44**). Young pairs enter new wood, primarily inhabiting pine or fir logs and stumps. Old colonies may consist of 3000 individuals or more and live in decayed logs. Occasionally structural wood is used, especially if damp. RANGE: Coastal and mountain areas. A similar species, the **Sierran Dampwood Termite** *(Z. nevadensis),* is more common in the northern half of the state, especially at higher and drier localities.

Family Rhinotermitidae

46, 47; Plate 1h. Western Subterranean Termite. *Reticulitermes hesperus.* This is the commonest and most destructive termite in California. ADULT: BL 5–6 mm; colonies consist of elongate-headed soldiers (**46**); 1–3 forms of reproductives; and blind, apterous, small (BL to 5 mm), whitish workers (**Plate 1h**); winged forms (**47**), BL 8–10 mm to tips of wings, blackish. Flights occur in the daytime, in spring or fall following the first rains. Mature colonies require 3–4 years to develop and are often massive, consisting of many thousands of individuals. RANGE: Throughout low to moderate elevations in California, wherever sufficient soil moisture and accessible wood are available. Damage to structural wood is common (primarily the understructure of houses), and fruit trees, grape vines, even potatoes are sometimes invaded.

Family Kalotermitidae

48, 49. Western Drywood Termite. *Kalotermes minor.* This is the commonest termite in coastal southern California. ADULT: Colonies consist primarily of whitish nymphs (BL to 8 mm); less common, red-brown to blackish soldiers (**48**) (BL 8–12 mm). In summer and fall numerous winged reproductives (**49**) appear with lustrous reddish or bluish black wings (L 11–12 mm to tips of wings). RANGE: Coastal areas and inland canyons from Mendocino southward. All kinds of sound wood are invaded including structural timbers (any portion of houses), dead tree branches, driftwood and, in drier, marginal portions of the state, buried portions of stumps and posts.

50. Desert Dampwood Termite. *Paraneotermes simplici-cornis.* Although a member of the Kalotermitidae, members of which characteristically use dry wood, this species is confined to moist wood in or on the soil in desert areas. It penetrates the earth at times but does not build mud-covered runways to reach wood above the ground. ADULT: Soldiers (**50**) are brown or yellowish brown with flat heads and short mandibles (BL 10–14 mm); nymphs are small (BL to 8 mm), whitish with the pronotum narrower than the head; and the reproductive alates are dark brown including the wings (L 13–15 mm to tips of wings). RANGE: Vicinity of Barstow eastward, usually in washes and sand dunes in association with dead roots or buried branches of mesquite and other desert shrubs, or sometimes structural wood in irrigated areas.

ROCK CRAWLERS
Order Grylloblattodea

Grylloblattids are cricketlike insects, intermediate in many respects between true Orthoptera (grasshoppers and crickets) and the mantids and walking sticks. Members of this order are wingless, slender, with two elongate tails or cerci, and chewing mouthparts. They are believed to be general feeders, eating soft-bodied insects or plant material.

Rock crawlers are found in cold, dark habitats such as under rocks, in decaying logs near receding snowbanks, and in ice caves. They are nocturnal and have been observed crawling on the snow in midwinter in subfreezing temperatures. Almost all collections of grylloblattids are made between October and May, and it is believed that these insects pass the warmer months secreted deep in cracks or rock slides. Their optimum temperature range is a few degrees above to a few degrees below freezing, and individuals quickly die if exposed to temperatures of 15°C or higher. The order is confined to Japan, Siberia, and western North America. There are about a dozen species in the western United States and Canada, 6 of which were described from California.

Reference

Gurney, A. B. 1948. Taxonomy and distribution of Grylloblattidae. *Proc. Ent. Soc. Washington* 50:86–102.

51. Ice crawlers. *Grylloblatta.* ADULTS: Body brownish, elongate (BL 25–30 mm); cylindrical; compound eyes small or absent; legs similar to one another; female with a long, sword-shaped ovipositor. RANGE: The similar-appearing California species are restricted to the Sierra Nevada, Cascade Range, and Modoc Plateau.

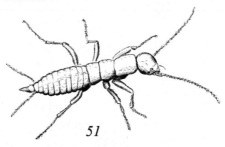

Fig. 51 Ice Crawler, *Grylloblatta* species.

CRICKETS, GRASSHOPPERS, KATYDIDS
Order Orthoptera

The order Orthoptera is a diverse group that includes both winged and wingless, heavy-bodied insects, most of which are adapted for jumping. Those that are winged possess long, straight, many-veined, stiff or leathery forewings and large, fan-shaped hindwings. All have chewing mouthparts and most feed on plants, although they sometimes prey on other insects or are scavengers of dead organic matter. There are more than 300 species in California, including many of our most familiar insects. Many are of economic importance to cultivated plants.

Well-developed muscles in the enlarged femora of the hind-legs give most members of this group exceptional jumping abilities. Many species are capable of producing sounds by rubbing together specialized frictional areas, either at the bases of the forewings (katydids and crickets) or between the hindleg and the forewing (grasshoppers). These songs usually function in courtship behavior, and in most species are produced by only the males. Both sexes possess hearing (tympanic) organs, located at the base of the foreleg tibia (katydids and crickets) or at the base of the abdomen (grasshoppers).

References

Helfer, J. 1963. *How to Know the Grasshoppers, Cockroaches and Their Allies*. Dubuque, Iowa: W. C. Brown; 353 pp.

Strohecker, H. F., W. W. Middlekauff, and D. C. Rentz. 1968. The grasshoppers of California (Orthoptera: Acridoidea). *Calif. Insect Survey Bull.* 10;177 pp.

GRASSHOPPERS
Family Acrididae

Grasshoppers differ from the majority of other jumping Orthoptera by their short, robust antennae, which seldom extend beyond the thoracic region. The hearing organs are large and located on the sides of the first abdominal segment. Although most adult grasshoppers are winged and many are strong fliers, some species are flightless. "Lubbers" (or "land lubbers") are short-winged, bulky forms which often are brightly colored, advertising their distastefulness to vertebrate predators.

The family Acrididae is large and diverse, with nearly 200 species known in California. Virtually all habitats, from seacoast and deserts to Alpine Fell-fields above timberline, are occupied by different species.

SPINE-BREASTED LOCUSTS
Subfamily Cyrtacanthacridinae

52. Gray Bird Grasshopper. *Schistocerca nitens*. ADULT: Both sexes fully winged; male (BL 40–50 mm to tips of wings) considerably smaller than female (BL 60–70 mm); gray or brownish; HW uniform translucent olive; a pale mid-dorsal stripe on head and prothoracic shield. NYMPH: Bright green. Adults, which are active in the spring, and immatures feed on a wide variety of garden and crop plants. RANGE: Widely distributed, from Monterey and Fresno counties southward.

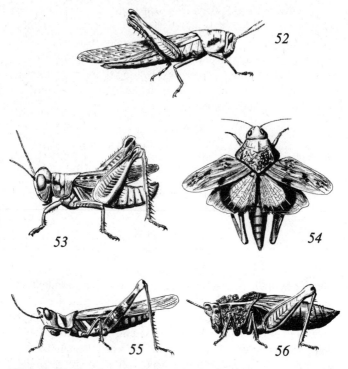

Figs. 52–56 Grasshoppers: 52, Gray Bird Grasshopper, *Schistocerca nitens;* 53, Devastating Grasshopper, *Melanoplus devastator;* 54, Lubberly Band-Winged Grasshopper, *Agymnastus ingens;* 55, Creosote Bush Grasshopper, *Bootettix punctatus;* 56, Dragon Lubber, *Dracotettix monstrosus.*

53. Devastating Grasshopper. *Melanoplus devastator.* This is the most widespread, abundant, and destructive grasshopper in the state. It is a member of a large genus (about 50 species in California). Because of its migratory and gregarious habits, *M. devastator* is a true locust and has been responsible for spectacular plagues during the history of California agriculture. ADULT: Pale to dark gray or brown; BL 20–24 mm; hind femur with three dark bars, tibia blue (rarely red); fully winged, HW clear, lightly tinged with greenish gray. Feeds on rangeland herbs and grasses and, during outbreak years, moves from the foothills into agricultural lands. RANGE: Typically found in semi-arid, grassy foothills from near sea level to 1500 m (5,000 ft) elevation.

BAND-WINGED GRASSHOPPERS
Subfamily Oedopodinae

The subfamily Oedopodinae has a larger number of species in California than any other subfamily, with many endemic species. Members of this subfamily commonly have brightly colored hindwings, often with a black or contrastingly colored band across the outer portion. The face is vertical or rounded rather than slanted, and the pronotum usually has a median ridge. The forewings are cryptically colored, like rocks, gravel, or sand, and camouflage the resting insect (**Plate 2a**).

Band-winged grasshoppers are widespread and most abundant in arid parts of the state, such as deserts, dry canyons, and sparsely vegetated, grassy mountainous areas. Few species are of appreciable economic concern. Their presence is most often noticed by their flight, which shows off the conspicuous hindwing color, and sometimes produces a crackling noise which is audible for considerable distances. The sound is created by the males snapping their wings together.

Plate 2a. Pallid-winged Grasshopper. *Trimeritropis pallidipennis*. One of our most widely distributed and common species, in a wide variety of habitats, up to 2600 m (8500 ft) elevation. ADULT: BL 31–42 mm to wing tips; FW dull tan or brownish with darker markings; HW pale yellowish or greenish with a black median band. The widespread **Red-winged Grasshopper** *(Dissosteira pictipennis)* is distinguished by having rosy-red HW on basal ⅓, followed by a broad, black band, and apical ⅓ colorless (BL 22–38 mm to wing tips). The closely related **Carolina** or **Mourning Cloak Grasshopper** *(D. carolina)* is spectacular, being much larger (BL 42–55 mm to wing tips), with pale tan to brown, unspotted FW, and HW colored like the Mourning Cloak butterfly (see **314**), bordered by a cream- or greenish-yellow band; males produce a loud, intermittent snapping while in flight; occurrence in California limited to the foothills and mountains from Mendocino County northward.

Plate 2b. Orange-winged or **Yellow-winged Grasshopper.** *Arphia conspersa*. This is a smallish (BL 22–40 mm to wing tips), springtime grasshopper with brightly colored hindwings, having a dark brown, outer band covering the apex. There are differing geographical forms: (**Plate 2b**) the **Orange-winged Grasshopper** *(A. conspersa ramona)* is common throughout southern California west of the deserts, north in the Coast Range to Mt. Diablo; HW bright orange. It is replaced in the southern Sierra Nevada and San Francisco Bay area northward by *A. conspersa behrensi,* which differs by having the HW greenish yellow or ochreous. The **Bloody-winged Grasshopper** *(A. pseudonietana)* occurs over the same northern range, flying late in the season, and is similar but has the HW bright blood-red. The **Robust Blue-winged Grasshopper** *(Leprus intermedius)* is a conspicuous species in southern California mountains and coastal areas during summer; HW bright blue with a broad, black band across the outer half; FW conspicuously marked with white and blackish crossbands; body robust (BL 32–46 mm to wing tips). **Sierran Blue-winged Grasshopper** *(Circotettix thalassinus)* is a familiar insect on high Sierran trails, occurring in the Sierra Nevada and Cascade Range at moderate to high elevations, and on the east side. This species is recognized by translucent, pale blue or greenish blue HW without a dark crossband, and by the dull, brownish gray, mottled FW, resembling the granitic surfaces *Circotettix* frequents; BL 32–40 mm to wing tips. Males fly high and hover over slopes where females are likely resting, issuing a high-pitched crackling which sometimes is mistaken for rattlesnake buzz at a distance. The **Clear-winged Grasshopper** *(Camnula pellucida)* is one of our most abundant species, occurring in grasslands of mountain meadows (to elevations above 2200 m (7000 ft), foothills, and valleys throughout the state. When abundant, adults and nymphs often migrate into agricultural areas; HW clear, colorless; FW brownish, mottled, with a conspicuous yellowish line, which meets to form a narrow ''V'' when the insect is at rest.

54. Lubberly Band-winged Grasshopper. *Agymnastus ingens.* ADULT: HW of both sexes yellow or greenish yellow with a broad, black marginal band; inner side of the broad hind femur blue toward the base; hind tibiae bright reddish orange. Female characterized by a very broad and rough prothoracic

shield; sluggish, grotesquely robust, and short-winged (BL 36–48 mm). Male long-winged and slender (BL 28–32 mm to tips of wings); normally active; often observed riding on back of female. RANGE: Rocky hillsides and mountain ridges in northern half of California.

SLANT-FACED GRASSHOPPERS
Subfamily Acridinae

55. Creosote Bush Grasshopper. *Bootettix punctatus.* ADULT: Medium-sized (BL 20–26 mm), slender grasshopper; head oblong, with an acute apex and slanted face; greenish with pearly white, oblique lateral areas on sides of thorax. A heat-loving species found clinging to twigs of Creosote Bush *(Larrea tridentata)* throughout the summer. The markings and colors of this insect, which serve effectively to hide it on the plant, are also exhibited by a slender katydid **(59)** *Insara covil-leae* (Tettigoniidae), which also lives on Creosote Bush. RANGE: High and low deserts of southern California. A less spectacularly colored, slant-faced acridid, *Ligurotettix coquil-letti,* which is often found on the same plant, perhaps is more commonly seen. It is a dull olive-gray species which stridulates from *Larrea* all day and on warm nights during late summer and fall. Its range is the same as that of *Bootettix.*

LUBBERS
Subfamily Romaleinae

56. Dragon Lubber. *Dracotettix monstrosus.* ADULT: Mottled gray and brown; very short wings in both sexes; a jagged median crest along back of prothoracic shield; female (BL 38–48 mm) much larger than male (BL 20–24 mm). Usually seen on gravelly soil amid grasses or shrubs on which it customarily feeds. Males active but females sluggish, relying on camouflage for protection. RANGE: West central and southern California; foothills and mountains.

KATYDIDS
Family Tettigoniidae

Katydids are large, leaf-mimicking insects with slender legs and very long, threadlike antennae (often exceeding the length of the body). By abrasion of special roughened files at the bases of the forewings, they produce a variety of notes. The sounds play an important part in courtship. (The note produced by one eastern species sounds like "Kate-she-did" and gave rise to a legend of an unhappy love affair. The phrase stuck as the common name of the family.) Small apertures in a swelling near the base of the tibia of the front leg mark the openings of the hearing organs.

57. Angular Winged Katydid. *Microcentrum rhombifolium.* ADULT: Large (BL to wing tip 50–65 mm); FW much broader at middle than at either end (rhomboid shape), giving the back of the insect an angular hump. Wings green, a perfect model of a leaf, complete with veins. SONG: Male (no bladelike ovipositor) emits short, lisping and ticking sounds which are

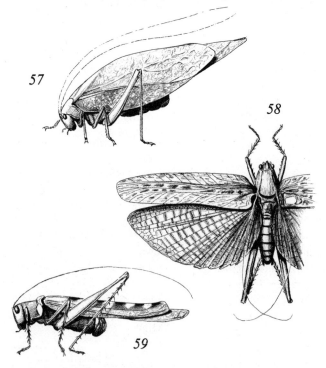

Figs. 57–59 Katydids: 57, Angular Winged Katydid, *Microcentrum rhombifolium;* 58, Brown-Winged Shield-Back, *Capnobotes fuliginosus;* 59, Creosote Bush Katydid, *Insara covilleae.*

responded to by low-intensity ticking from female (curved bladelike ovipositor). Calls from trees in late summer and fall at dusk and through the night when warm. EGGS: Flat, oval, gray, placed by female on twigs in a single or double overlapping row. RANGE: Common throughout the state.

Plate 2c. Fork-tailed Bush Katydids. *Scudderia*. ADULTS: Smaller than preceding species (BL to wing tips 30–38 mm); wings slender and straight. SONG: Male usually lisps ("zeek" or "zip") in series of 3 or 4; female may respond with a ticking noise. Habits otherwise similar to those of previous species. EGGS: Flat and inserted lengthwise into the edges of leaves. NYMPHS: Distinguishable from that of the Angular Winged Katydid by a horn on the head between the antennae. RANGE: Widespread in the state; *S. mexicana* (**Plate 2c**) is common in southern California, *S. furcata* is more common in the northern part.

Plate 2d. Splendid Shield-back Katydid. *Neduba ovata*. This is California's most characteristic representative of a group of katydids which are dull colored (mimicking sticks or wood), with abbreviated wings and enlarged thoracic shields. ADULT: Large (BL 25–40 mm); body yellowish, with brown and greenish mottling; wings completely hidden beneath the thoracic shield. Active in late summer and fall. SONG: Males repeat a sharp "zip-zip-zip" series between long pauses. RANGE: Typical form in desert portion of southern California. Other races are found in the state, including a larger one along the eastern slopes of the Sierra Nevada.

58. Brown-winged Shield-back. *Capnobotes fuliginosus*. This is a spectacular desert insect because of its large size (BL to wing tips up to 75 mm) and ear-splitting call. ADULT: Fully winged; brownish gray, mottled, HW darker than FW. Found all summer on Mesquite and other shrubs in desert; most active on hot nights. Partly carnivorous. SONG: Very loud, continuous, shrill buzz. RANGE: Desert portions of southern California.

59. Creosote Bush Katydid. *Insara covilleae*. ADULT: Slender (BL to wing tips 28–30 mm); color basically green, forewings marked with a series of oblique, light green bars; pearly blotches on sides of the prothoracic shield. This is a graceful insect which lives on Creosote Bush *(Larrea tridentata)* in the desert. Its broken pattern and green color provide camouflage amid the dissected leaves of the plant (similar in color to the grasshopper, **(55)** *Bootettix*). RANGE: Restricted to desert areas of southeastern parts of the state.

GROUND AND CAMEL CRICKETS
Family Gryllacrididae

Most gryllacridids live on the ground or in burrows. All are pale or dull colored and resemble katydids in having long antennae, legs, and body form. They differ by lacking wings and are silent. The auditory organs on the forelegs are missing. Although most species are ground-dwellers, a few occur in trees. In California, species of *Gammarotettix*, **chaparral camel crickets,** are brownish gryllacridids commonly found in shrubs, oaks, and other trees in foothill areas.

60. California Camel Cricket. *Ceuthophilus californianus.* ADULT: Body arched or "hump-backed" (BL 15–25 mm); yellowish to reddish brown; hind tibia of both sexes stout and straight. Common in moist grasslands where it lives in underground burrows, gopher holes, under stones, boards, etc. RANGE: General, at lower and middle elevations.

61. Sand treaders. *Macrobaenetes.* ADULTS: Have large bristles on the hind tibiae that form a basket for pushing sand, adapting them for a burrowing life on desert and coastal dunes.

62. Mushroom camel crickets. *Pristoceuthophilus.* Males have bowed hind tibiae and roughened or tuberculate areas on the back of the abdomen. They are destructive sometimes in commercial mushroom plantings.

Family Stenopelmatidae

Plate 2e. Jerusalem Cricket or **Potato Bug.** *Stenopelmatus fuscus.* No other California insect inspires such awe. Its large size (BL up to 50 mm), spiny legs, giant jaws and, more than anything, its oversized, foreboding, humanoid, bald head, give

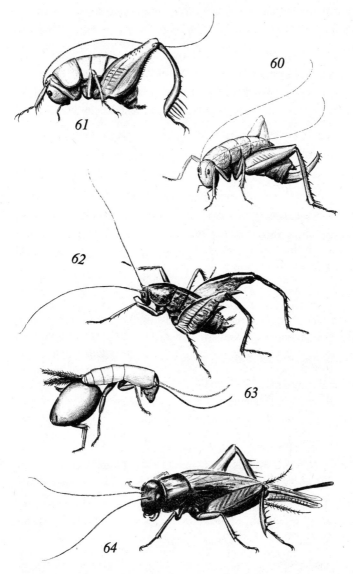

Figs. 60–64 Crickets: 60, California Camel Cricket, *Ceuthophilus californianus;* 61, Sand Treader, *Macrobaenetes* species; 62, Mushroom Camel Cricket, *Pristoceuthophilus* species; 63, Ant Cricket, *Myrmecophila oregonensis;* 64, Field Cricket, *Gryllus* species.

rise to fear that it is dangerous and even a demon. Also known as Woh-tzi-Neh ("Old Bald-Headed Man") and "Niña de la Tierra" ("Child of the Earth"). It is not poisonous and is a harmless native insect, most commonly encountered by gardeners when turning the soil in the spring. ADULT: Generally tan; abdomen bulbous with black segmental rings; wingless; eyes small. Burrows in soil, feeding on roots and tubers. May wander above ground at night. RANGE: Widespread in state. There are several other similar species in California. This genus is restricted to western North America.

CRICKETS
Family Gryllidae

Crickets are the virtuosos among insect singers. Their melodious trilling and chirping, part of courtship, are produced like the notes of katydids, by stridulation of roughened veins at the bases of the forewings. In crickets only the males sing.

Plate 2f. Tree crickets. *Oecanthus*. Tree crickets are quite different from field crickets; instead of having a cylindrical, robust, heavy, darkly pigmented form, their bodies are frail, pale greenish or whitish, translucent; the males, with their greatly expanded wings, appear rather flattened. There are several species, distinguishable only by details of markings and song. ADULTS: Small (BL 10–15 mm); male broad, with flattened, transparent wings; female cylindrical, wings folded down sides of body. SONGS: Varying series of rolling chirps, "treet-treet-treet" or continuous, soft trill. Usually call from shrubs or trees on warm nights. Because the pitch and frequency of chirps are a function of temperature, *Oecanthus* are also called "thermometer crickets." RANGE: General.

63. Ant Cricket. *Myrmecophila oregonensis*. This is one of a variety of insects whose lives are spent as "guests" in social insect colonies. ADULT: Wingless, small (BL 2–3 mm); brown. Found in and near the entrance of the nests of various species of ants—under bark of stumps and logs or under boards and rocks on the ground. Apparently it feeds on the host's food and secretions. RANGE: Widespread, especially in arid habitats.

64. Field crickets. *Gryllus*. Species of this genus are the common native crickets of fields and grasslands. ADULTS: Medium-sized (BL 15–30 mm); dark brown to black; usually solid colored, but sometimes with light markings on wings. SONG: Intermittent, shrill trill. During certain years population explosions of this species may occur, and plagues may cause heavy crop losses. During such times, following summer thunderstorms, thousands of crickets may congregate around lights at gas stations, restaurants, etc., and constitute a considerable nuisance. Normally, individuals live under objects on the ground and feed on native herbaceous plants. RANGE: Ubiquitous. The introduced **House Cricket** *(Acheta domestica)* is much rarer in California. It can maintain itself indoors and is established outdoors in a few southern California localities (Orange County; Calexico). This species is distinguished by its light brown color and banded head.

MANTIDS
Order Mantodea

Mantids are unique orthopteroid insects because of their predaceous habits. They are easily distinguished by the elongate prothorax and forelegs modified for grasping. They perch waiting for prey, with front legs upraised in a "praying" attitude. The eggs are laid in hardened froth cases attached to vegetation or rocks. Adults are primarily nocturnal and are readily attracted to light.

Only the first 2 of the illustrated species are California natives. The remainder have been introduced and have only local ranges near their points of release. These insects are sometimes used as biological control agents, such as in gardens, by introduction of the egg cases.

Reference

Gurney, A. 1951. Praying mantids of the United States, native and introduced. *Ann. Rept. Smithson. Inst.,* 1950; pp. 339–362.

65. California Mantid. *Stagmomantis californica*. ADULT: Moderately large (BL 50–65 mm); green, yellow, and brown color phases; wings of male extending well beyond tip of abdomen, short (only to middle of abdomen) in female; HW

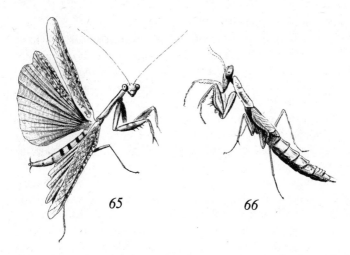

Figs. 65–66 Mantids: 65, California Mantid, *Stagmomantis californica;* 66, Minor Ground Mantid, *Litaneutria minor*.

generally dark, marked anteriorly with ashy blotches; abdomen with 4 conspicuous, dark, transverse bands. Males commonly come to light. RANGE: Essentially a desert and arid land species; found in all of southern California and north through Central Valley. The **Carolina Mantid** *(S. carolina)* is much rarer and of local occurrence. ADULT: Similar to the California Mantid, but generally smaller (BL 45–50 mm); more often green; abdomen lacks dark, basal crossbands; female abdomen strongly widened at middle. RANGE: We have verified collection records only from Modesto.

66. Minor Ground Mantid. *Litaneutria minor*. ADULT: Small (BL 20–30 mm); slender; pale grayish brown; male with fully developed wings; female with short or full-sized wings; HW has a dark purplish blotch near the base in both sexes. Very active and normally found on the ground in chaparral or coastal sage scrub. RANGE: A native species in sparsely vegetated, dry portions of state; deserts and San Diego north through the Central Valley and Sierran foothills.

Plate 2g. Mediterranean Mantid. *Iris oratoria*. ADULT: Medium-sized (BL approximately 50 mm); green and brown phases; both male and female with well-developed wings, the female's shorter, exposing about half the abdomen; HW with a shining bluish black blotch near the base and an area of brick red towards the front. RANGE: Introduced from the Mediterranean region; established in the Central Valley and southern California (San Joaquin, Coachella, and Imperial valleys) and apparently expanding its range in the state. Another introduced, even larger species, the **Chinese Mantid** *(Tenodera aridifolia)*, has established local populations in various parts of the state, usually in suburban areas. In the form of egg masses, this and *I. oratoria* are the 2 mantids most widely sold to gardeners for pest control. Escapees readily naturalize where climate permits. ADULT: Large (BL 75–100 mm); green; wings complete in both sexes and somewhat jointed apically; HW mottled throughout; abdomen without dark crossbands. RANGE: Native to the Orient. Known from the San Francisco Bay area, Los Angeles basin, and a few other locations.

WALKING STICKS
Order Phasmatodea

Walking Sticks are large insects that in California are wingless, elongate, and cylindrical. They possess well-developed mandibles for chewing, like those of jumping Orthoptera, and feed on plants. The legs are all similar and are widely spaced. The ovipositor is small and mostly concealed by an enlargement of the abdominal underside. Slow movement and the body form mimicking twigs make these insects difficult to find in nature. There are only a few species in California.

67. Western Short-horned Walking Stick. *Parabacillus hesperus.* ADULT: BL 60–90 mm; antenna much shorter than front femur; brown; smooth-bodied. Feeds on grasses and soft shrubs such as burroweed *(Haplopappus)* and desert mallow *(Sphaeralcea)*. RANGE: Known from widely scattered California localities; more abundant towards the south and in the mountains.

68. Gray Walking Stick. *Pseudosermyle straminea.* ADULT: BL 40–75 mm; antenna longer than front femur; usually light gray, occasionally pinkish or yellow; thorax spiny,

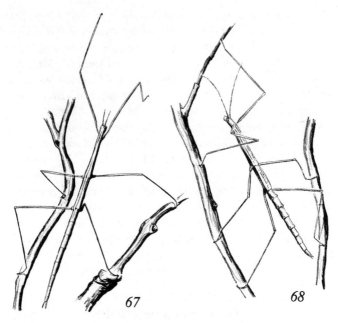

Figs. 67–68 Walking Sticks: 67, Western Short-Horned Walking Stick, *Parabacillus hesperus;* 68, Gray Walking Stick, *Pseudosermyle straminea.*

roughened or ridged lengthwise, especially in female. On grasses and various grayish desert shrubs. RANGE: Arid portions of southern California, at low elevations.

Plate 2h. Timemas. *Timema.* These are atypical walking sticks having stout bodies and short legs. ADULTS: Small (BL less than 35 mm); green, brown, or pinkish forms. Run actively when disturbed and secrete an unpleasant odor. Live on shrubs and trees where they may be abundant in the spring and early summer. RANGE: The several species in the state are characteristic members of the chaparral and oak woodland communities and occur throughout state in suitable habitats.

EARWIGS
Order Dermaptera

Earwigs are nocturnal, terrestrial insects that live in moist, secluded places, such as cracks in the soil and under boards or other refuse around man's habitations. They are somewhat flattened, elongate, with incomplete metamorphosis, a tough, shiny integument, and well-developed, movable forceps at the tip of the abdomen. The adults may be winged or wingless. When present, the forewings are short and leathery and cover all but the tips of the large, membranous hindwings which are folded radially when at rest. The large, spherical eggs are laid in clusters and are guarded by the mother. Of 10 species known in California, all but one are introduced from other parts of the world.

Reference

Langston, R. L., and J. A. Powell. 1975. The earwigs of California (Order Dermaptera). *Calif. Insect Survey, Bull.* 20; 25 pp.

Family Forficulidae

69. European Earwig. *Forficula auricularia.* Although this is the most abundant earwig in California, it was not known in the state until 1923. ADULT: Medium-sized (BL 12–22 mm); mostly brown with pale forewings and antennae; forceps much larger and more widely spaced in male. The immatures and adults feed on a wide variety of substances, from flowers and

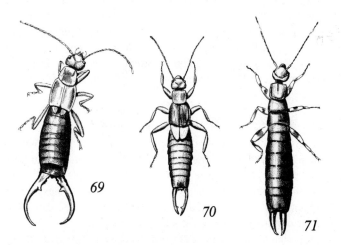

Figs. 69–71 Earwigs: 69, European Earwig, *Forficula auricularia;* 70, Toothed Earwig, *Spongovostox apicedentatus;* 71, Ring-Legged Earwig, *Euborellia annulipes.*

green foliage near the ground to living and dead insects, including aphids. RANGE: Los Angeles northward throughout low to moderate elevations in the state.

Family Labiidae

70. Toothed Earwig. *Spongovostox apicedentatus.* This species is the only native earwig in California. ADULT: Small (BL 8–12 mm); fully winged, reddish brown; forceps in male possess a conspicuous subapical tooth. Most commonly found in dead cactus, yucca, and rotting wood of other desert plants. RANGE: Southern California coastal areas, margins of the Mojave Desert, and in the low deserts.

Family Carcinophoridae

71. Ring-legged Earwig. *Euborellia annulipes.* ADULT: Medium-sized (BL 15–17 mm), wingless species having pale legs ringed with dark brown. RANGE: Established around Los Angeles at least since the 1880s and through the central part of the state since the turn of the century. A closely related species, the **African Earwig** *(E. cincticollis)* has spread rapidly through the arid parts of California, including the Central Valley, since its discovery at Blythe in 1946. ADULT: Small to medium-sized (BL 9–15 mm); dark brown with pale legs and dark antennae having the 3rd and 4th preapical segments pale. *E. cincticollis* is polymorphic, with adults either wingless, fully winged, or short-winged. Both species are general feeders and are the commonest earwings in southern California.

WEBSPINNERS
Order Embioptera

Webspinners are small, elongate, flattened insects that live in silken galleries among grass roots under rocks, in leaf litter, and similar habitats. Silk is produced from glands in the greatly swollen front legs. Females are wingless, while males in many species are fully winged. Embiids are easily recognized by their distinctive body form with widely spaced, short legs and by their ability to run backward. The wings are flexible, enabling reverse movement in the galleries, and the flight is weak and termitelike.

A great diversity of webspinners occurs in tropical regions. Only 3 species are known in California, 2 of them introduced from the Mediterranean region. All are scavengers, eating decayed plant matter.

References

Ross, E. S. 1944. A revision of the Embioptera, or webspinners, of the New World. *Proc. U.S. Natl. Museum* 94;104 pp.
———— 1957. The Embioptera of California. *Calif. Insect Survey, Bull.* 6(3):51–58.

Family Oligotomidae

72. Pink Webspinner. *Haploembia solieri.* This introduced species has been known in California from colonies of asexual females since before the turn of the century. Colonies are easily recognized by the extensive silk trackways running through

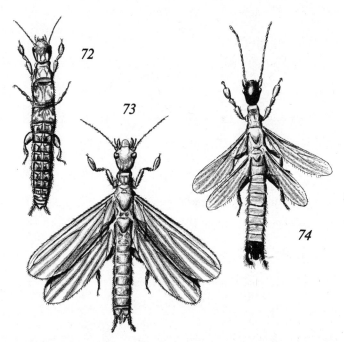

Figs. 72–74 Webspinners: 72, Pink Webspinner, *Haploembia solieri* (asexual female); 73, Black Webspinner, *Oligotoma nigra* (male): 74, Red Webspinner, *Chelicerca rubra* (male).

grass roots at the edges of stones, under boards, and in dry cow chips. Recently a bisexual colony was discovered at Redwood City, suggesting that a separate introduction occurred from one of the areas of the Old World where both sexes occur. ADULT: BL 6–10 mm; wingless, soft-bodied, pale with a faint, mottled pattern of reddish on the back that renders a pink appearance. JUVENILE: Similar, smaller. RANGE: The most widespread and commonly encountered embiid in California, inhabiting grassy areas from the northern end of the Central Valley southward, throughout foothill areas of the Sierra Nevada and the Coast Ranges.

73. Black Webspinner. *Oligotoma nigra.* The species is common around trunks of palms and in other habitats and is believed to have been introduced with date palm cuttings from Egypt in the 1890s. ADULT: BL 6–10 mm; female blackish brown; male uniformly pale brown to dark brown including wings; cerci at the tip of his abdomen 2-segmented, similar in form. JUVENILE: Tan to brown. RANGE: Widely established in deserts and coastal portions of southern California, where the males often are attracted to lights.

Family Anisembiidae

74. Red Webspinner. *Chelicerca rubra.* This is the only native embiid in California. Colonies usually occur under stones on well-drained, grassy slopes with cactus and other xeric plants. The males are believed to disperse in the daytime. ADULT: Small (BL 5–7 mm), red; male winged, with terminalia and head black; cerci at the tip of his abdomen dissimilar, the left one 1-segmented, curved. JUVENILE: Pale red-brown. RANGE: Near the coast in San Diego County, through desert margins north to Tehachapi.

BARK LICE
Order Psocoptera

The free-living insects in the order Psocoptera are called "lice" only because of their louselike appearance. Although small and soft-bodied, they frequently have wings, and none is parasitic; they live on tree trunks or foliage, under bark or stones, etc., where they feed on molds, pollen, and a variety of organic matter. A structural feature readily identifying the order is the swollen or bulbous area on the front of the head between the widely spaced antennae.

The California species are incompletely classified but probably at least 50 species are recorded.

75. Ten-spotted Psocid. *Peripsocus californicus.* This common species is a typical outdoor psocid. ADULT: BL 2 mm; wings transparent, with few veins and small, brown spots. Common on the leaves of many kinds of trees. Gregarious, groups of adults and young living under a frail web spun by the adult from glands opening on the lower lip. RANGE: Central California coast.

76. Cereal Psocid. *Liposcelis divinatorius.* This is one of several domestic members of the order. Individuals are common inhabitants of homes and buildings where they appear as minute brown specks moving over paper (''book lice''), window sills, wallpaper, or other surfaces. ADULT: Tiny (BL less than 1 mm); wingless; white or brown; hind femur enlarged. Feeds on and may damage organic matter of all kinds (cereal products, wallpaper and book binding pastes, dead insects), especially if slightly damp; it is not clear whether their primary food is molds on these materials or the substances themselves. RANGE: Urban areas.

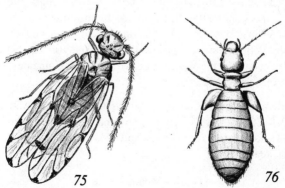

Figs. 75–76 Bark Lice: 75, Ten-Spotted Psocid, *Peripsocus californicus;* 76, Cereal Psocid, *Liposcelis divinatorius* (adult).

BITING LICE
Order Mallophaga

Members of the order Mallophaga are called biting lice or bird lice because they are external parasites of vertebrates, principally birds. Some forms make mammals their hosts, but never man. These wingless lice are distinguished from sucking lice (Anoplura) by possessing chewing, mandibulate mouthparts. Food of biting lice consists of feathers or fur and the outer layer of skin. Because most Mallophaga tend to be fairly host specific and we have an extensive avifauna, California has a large number of species (at least 170).

Family Gyropidae

77. Guinea Pig Louse. *Gliricola porchelli*. ADULT: Small (BL 1–1.2 mm); body slender; head triangular, antennae fitting into notches on sides; claws of legs greatly reduced; hind 2 pairs of legs with tibiae and femora curved and grooved, with opposing movement for clasping hairs. Remain close to skin, clinging to a single hair. HOST: Guinea Pig only, not found on native mammals. RANGE: Wherever Guinea Pigs are kept in captivity.

Family Philopteridae

78. Pelican Pouch Louse. *Piagetiella bursaepelecani*. ADULT: Large (BL 4–5 mm); body elongate; dark brown, nearly black laterally; head broad, hiding small antennae. HOST: This louse inhabits the inside of the pouch of the Brown Pelican and is apparently host specific (a distinct species occurs on the White Pelican). RANGE: Coincides with range of host, entire coast.

79. Slender Pigeon Louse. *Columbicola columbae*. ADULT: BL 2.1–2.6 mm; very slender; antennae large and exposed. HOST: Restricted to pigeons and related birds. RANGE: Widespread.

Family Menoponidae

80. Chicken Body Louse. *Menacanthus stramineus*. ADULT: BL 2–2.5 mm; slender with broad abdomen; head

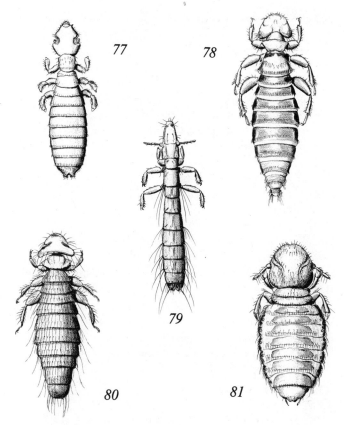

Figs. 77–81 Biting Lice: 77, Guinea Pig Louse, *Gliricola porchelli;* 78, Pelican Pouch Louse, *Piagetiella bursaepelecani;* 79, Slender Pigeon Louse, *Columbicola columbae;* 80, Chicken Body Louse, *Menacanthus stramineus;* 81, Livestock Biting Louse, *Bovicola* species.

triangular; antennae small and hidden; abdominal segments with two transverse rows of hairs; straw yellow (dark spots from food in intestine). Often develops very large populations (35,000 on a single bird). Favors portions of skin not densely feathered, especially below vent. HOST: Common and most injurious louse on chickens and other poultry species. RANGE: Widespread, wherever chickens or turkeys are housed. Although these lice do not suck blood, they result in a nervous condition of infested birds which prevents sleep, causes loss of appetite, emaciation, and reduction in number of eggs laid. Young birds brooded by lousy hens are often killed by lice swarming to them.

Family Trichodectidae

81. Livestock biting lice. *Bovicola.* Several species in this genus parasitize mammals, including domestic livestock. When excessively abundant they may cause a kind of mange and severe irritation to the host's skin. ADULTS: Small (BL 1–2 mm); head circular in outline; antennae large and exposed; abdomen with distinct transverse bands. HOSTS: Cattle, horses, goats, dogs, deer, etc. RANGE: Widespread, wherever livestock are raised.

SUCKING LICE
Order Anoplura

The order Anoplura is distinguished from the biting or bird lice by mouthparts formed of long, needlelike stylets adapted for sucking blood. Sucking lice are wingless, external parasites of mammals including man, who harbors two species all his own. Anoplura has fewer species (around 40 in the state) than the Mallophaga because of the smaller number of hosts.

References

Keh, B., and J. Poorbaugh. 1971. Understanding and treating infestations of lice on humans. *Calif. Vector Views* 18:24–31.

Ferris, G. F. 1951. The sucking lice. *Pacific Coast Ent. Soc. Memoir* 1; 320 pp.

82. Human Louse. *Pediculus humanus.* This insect, also commonly called the "cootie," is a serious enemy of man. It acts as a vector of epidemic typhus and has caused disease and death to millions of persons in historical times. ADULT: Small (BL 1.5–2 mm); grayish white; elongate oval with greatly swollen abdomen in blooded specimens. There are two forms recognized: the **Head Louse** *(P. humanus capitis)* infests the head and tends to be smaller and with proportionately heavier legs than the **(82) Body Louse** *(P. humanus humanus),* which primarily locates itself under the clothing on the body. EGGS ("nits"): White, attached firmly to the hairs. HOST: Man.

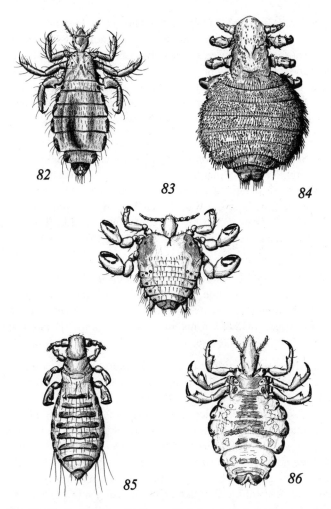

Figs. 82–86 Sucking Lice: 82, Human Body Louse, *Pediculus humanus;* 83, Crab Louse, *Pthirus pubis;* 84, Seal Louse, *Echinophthirius horridus;* 85, Spiny Rat Louse, *Polyplax spinulosa;* 86, Livestock Sucking Louse, *Haematopinus* species.

RANGE: Usually limited locally to persons practicing poor hygiene, but frequently breaks out among school children.

83. Crab Louse. *Pthirus pubis.* This nuisance (not known to transmit diseases) inhabits the hairs of the pubic region almost entirely. It is short and broad, having large legs with very heavy claws (crablike), features which readily distinguish it from the Human Louse. ADULT: BL 1.5–2 mm; body triangular, abdomen tapering posteriorly with short side lobes. HOST: Known only to infest white and Negro races of man. Transmitted by close contact, most commonly during sexual intercourse. RANGE: Wherever poor hygiene accompanies sexual promiscuity.

84. Seal Louse. *Echinophthirius horridus.* Several kinds of sucking lice infest marine mammals off our coast. Valves which close breathing pores prevent the lice from drowning when the host is in the water. ADULT: BL 3.5 mm; heavily pigmented, brown; densely covered with scalelike bristles. HOST: Seals, especially Harbor Seal. RANGE: Congruent with hosts' range.

85. Spiny Rat Louse. *Polyplax spinulosa.* Small, slender (BL 1 mm); pale white, with conspicuous, transverse back plates bearing large bristles; antennae arising near anterior end of the head. HOST: On native as well as domestic rats and mice. RANGE: Widespread; wherever rats breed.

86. Livestock sucking lice. *Haematopinus, Linognathus.* These genera, in parallel to (**81**) livestock biting lice, form a complex of species on domestic mammals. ADULTS: Large (BL 2–6 mm); head elongate with protuberance behind the antenna (**86**) *(Haematopinus)* or short and without head lobe *(Linognathus);* abdomen of latter also covered with dense rows of long bristles as opposed to only scattered bristles on former. HOSTS: Cattle, horses, hogs, dogs, etc. Locate mostly in the ears and folds of the skin, sometimes causing severe irritation. RANGE: Wherever hosts occur.

TRUE BUGS, HOPPERS, APHIDS, SCALES, AND OTHERS
Order Hemiptera

The Hemiptera is the fifth largest order of insects, and it includes a great diversity of forms. All possess a tubular beak adapted for piercing and sucking. Most have 2 pairs of wings in the adult, while some are seasonally or always wingless. There are 2 suborders: the Heteroptera (true bugs) in which the apical part of the forewing is membranous and the remainder usually thicker and leathery; and the Homoptera (cicadas, hoppers, scales, aphids) in which the forewing is uniform in consistency, often quite similar to the hindwing. The metamorphosis is gradual; nymphs resemble the adults except for lacking wings and they pass through usually 4 or 5 instars. The last instar in male scale insects and whiteflies is immobile. It is often called the "pupa," owing to its resemblance to the pupa of holometabolous insects. The California fauna probably contains in excess of 2,000 described species, but there is no accurate estimate.

TRUE BUGS
Suborder Heteroptera

The true bugs are most reliably recognized by the form of the wings, although adults of a few species are short-winged or wingless. The forewings are thickened, often brightly colored, leathery in consistency, with membranous tips; the hindwings are membranous, fanlike, and fold under the forewings when at rest. When the wings are folded, the membranous portions of the forewings overlap, and with the dissimilar basal parts, form a characteristic "X" on the back, by which nearly all Heteroptera can be recognized.

Many true bugs are aquatic or semi-aquatic, living in the water, on its surface, or at wet margins of ponds and streams; practically all of these are predaceous on other small animals. The great majority of species are terrestrial, living on plants, where most are herbivorous and a few are predaceous. A very few are bloodsucking parasites of mammals or birds.

The suborder Heteroptera includes species exhibiting a great diversity of size, color, structural modifications, and habitat preferences. More than 600 species are known in California.

References

DeCoursey, R. M. 1971. Keys to the families and subfamilies of the nymphs of North American Hemiptera-Homoptera. *Proc. Ent. Soc. Washington* 73:413–428.

Menke, A. S. (ed.), et al. 1979. The semiaquatic and aquatic Hemiptera of California. *Calif. Insect Survey, Bull.* 21.

Miller, N.C.F. 1956. Biology of the Heteroptera. London: Methuen Co.; 172 pp. (Reprinted 1971 by Ent. Reprint Specialists, Los Angeles).

Usinger, R. L. 1966. *Monograph of Cimicidae (Hemiptera, Heteroptera).* College Park, Md.: Thos. Say Foundation; 585 pp.

STINK BUGS
Family Pentatomidae

Pentatomids are among the most easily recognized true bugs because of their fairly consistent 5-sided outline. The scutellum is large, extending about half the length of the wings, and is narrowed apically. Most species are moderately large. They may not be easily seen on plants, but their presence is often noticed by a characteristic, powerful odor from a fluid ejected from ventral thoracic glands when the bugs are disturbed. These bugs are fairly strong fliers but primarily rely on cryptic coloration and sedentary behavior to escape detection. The majority are plant feeders, but species of one subfamily are predaceous on other insects. About 50 species occur in California, representing a diversity of colors, forms, and habitats.

87. Green stink bugs. *Chlorochroa.* Two species of this genus are among the largest and most often seen of California's pentatomids. (**87**) **Say's Stink Bug** *(C. sayi).* ADULT: BL 10–15 mm; bright to dark green, with variable amounts of white speckling on the back, the tip of the scutellum, and three spots at its base, pale yellow or white. NYMPH: Variable, pale to dark green with orange markings. The bugs are found on a wide variety of plants, including wheat, alfalfa, and other field crops, where their feeding sometimes causes damage. RANGE: East of the Sierra Nevada, desert areas, coastal southern California, and Central Valley. The **Conchuela** *(C. ligata)* is similar in appearance and range but is larger (BL 12–16 mm) and dark olive green with a narrow marginal border and the tip of scutellum red or orange. The nymphs are gray with reddish markings and are often gregarious, living in compact colonies. The species is sometimes injurious to a variety of crops.

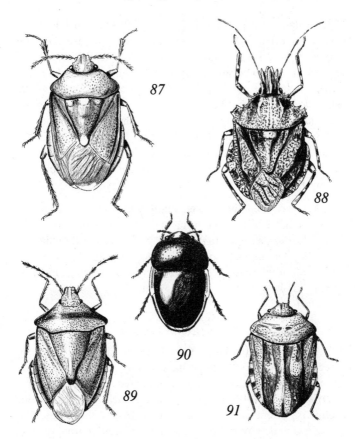

Figs. 87–91 True Bugs: 87, Say's Stink Bug, *Chlorochroa sayi;* 88, Rough Stink Bug, *Brochymena* species; 89, Western Red-shouldered Plant Bug, *Thyanta pallidovirens;* 90, Negro Bug, *Corimelaena* species; 91, Shield Bug, *Eurygaster* species.

Plate 3a. Harlequin Cabbage Bug. *Murgantia histrionica.* ADULT: BL 7–11 mm; brightly colored, black with variable orange or reddish markings. EGGS: White with conspicuous black rings; deposited in clusters, resembling miniature white barrels with black hoops. NYMPH: Shining black with conspicuous white, yellow, and orange markings. All stages live on plants of the mustard family and may be found throughout the year in warmer parts of the state. RANGE: Lower elevation areas throughout California.

88. Rough stink bugs. *Brochymena*. Members of this genus are most often found on trees and shrubs where they are predaceous on soft-bodied insects. ADULTS: Large (BL 12–18 mm); brownish or gray with a roughened surface and strongly angulate outline. RANGE: Throughout most of the state, at low to moderate elevations. The Rough Stink Bug *(B. sulcata)* is the most common species in the Sierra Nevada, the Central Valley, and the Coast Ranges. It often occurs in orchards of central and southern California, where it is regarded as a beneficial insect because both nymphs and adults prey on caterpillars and other pests.

89. Western Red-shouldered Plant Bug. *Thyanta pallidovirens*. ADULT: BL 8–11 mm; variable, brownish to bright green, often with a distinct red band across the thorax. RANGE: Throughout most of the state, up to 1800 m (6000 ft) elevation in the Sierra Nevada. Common on a wide variety of grains, flowers, field crops, and shrubs.

NEGRO BUGS
Family Corimelaenidae

90. Common negro bugs. *Corimelaena*. These bugs usually feed on plants of the nightshade family, including tomato, where they are occasionally encountered in large clusters. ADULTS: Small (BL 3–4 mm); shining black, sometimes with yellowish or orange lateral marks at the base of the wings. The back is almost covered by the greatly enlarged, strongly convex scutellum. RANGE: Throughout lower and moderate elevations of the state except the deserts.

SHIELD BUGS
Family Scutelleridae

Shield bugs differ from stink bugs in having the scutellum of the thorax greatly enlarged to form a broad, rounded shield that almost entirely covers the back and wings. About a dozen species are known in California. They are widespread and common in weedy fields but are often overlooked owing to their cryptic coloration and behavior.

91. Common shield bugs. *Eurygaster*. ADULTS: BL 7–12 mm; variable in color and markings; pale to dark brown, usually with distinct, longitudinal or triangular, pale markings on the shield. Most commonly encountered in late spring when the bugs climb up onto drying grasses where they are often collected in sweep nets. RANGE: Coast Ranges and Sierra Nevada foothills, to about 2000 m (6500 ft) elevation.

LEAF-FOOTED BUGS
Family Coreidae

Coreids are often brightly colored, with enlarged, flat, leaf-like hind legs. They feed by piercing plant parts with their elongate beaks and sucking out the juices. Frequently they select fruits and are sometimes destructive to cultivated plants. More than 40 species are known in California, mostly from warmer parts of the state.

92. Squash Bug. *Anasa tristis.* This is a widespread, common, and often destructive insect. ADULT: BL 12–17 mm; pale or dark brown with the margins of the abdomen that protrude beyond the wings alternately striped with orange or tan and brown. NYMPH: Pale green with pinkish appendages and often covered with a whitish powder. Later instars have dark appendages. All stages feed on cucurbits, especially melons and squashes, but also wild gourds and manroot. RANGE: Coastal areas, desert margins, Central Valley, and Sierran foothills.

Plate 3b. Western Leaf-footed Bug. *Leptoglossus clypealis.* ADULT: Large (BL 16–19 mm); uniform brown with an irregular, narrow, white band across the middle of the wings, yellow marginal spots on the abdomen, and leaflike enlargements of the hind legs. RANGE: Wide variety of habitats in lower elevation deserts and Central Valley of California; common on juniper and sometimes injures young green fruit in orchards and cotton fields. Occasionally enters homes during cold weather.

Family Rhopalidae

Plate 3c. Western Box Elder Bug. *Leptocoris rubrolineatus.* This species is especially noticed in fall and winter when the adults often migrate into buildings for hibernation, and in early spring when large numbers emerge on the first warm days

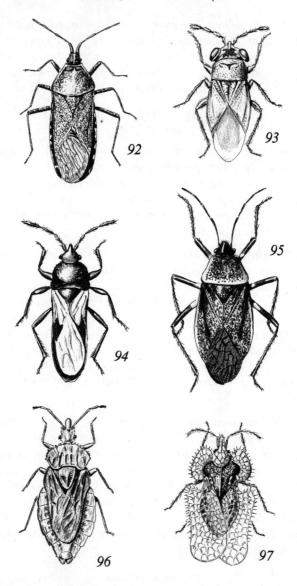

Figs. 92–97 True Bugs: 92, Squash Bug, *Anasa tristis;* 93, Big-Eyed Bug, *Geocoris* species; 94, Southern Chinch Bug, *Blissus insularis;* 95, Bordered Plant Bug, *Largus cinctus;* 96, Flat Bug, *Aradus* species; 97, Lace Bug, *Corythucha* species.

and swarm onto all kinds of plants. ADULT: Elongate (BL 9–13 mm); rather flat; gray to black with conspicuous red lines on the thorax and wings; abdomen bright red and conspicuous when the bugs are in flight. NYMPH: Gray and bright red, becoming marked with black in later stages. All stages normally feed on Box Elder and maple, but sometimes they go to fruit trees and puncture developing fruit. RANGE: Widely distributed in California north of the deserts, at low to moderate elevations.

Family Lygaeidae

Most lygaeids are dark-colored bugs, often with short wings, that live on and under plants and in other secluded spots where they feed on fallen seeds. Some species feed on green plant parts, and a few are brightly colored, warning predators of distastefulness acquired from milkweeds upon which they feed. More than 70 species have been recorded in California.

Plate 3d. Large Milkweed Bug. *Oncopeltus fasciatus*. This bug is often used in laboratory experiments because it is easily reared in large colonies. In nature the adults and nymphs usually are restricted to milkweed, sometimes becoming abundant, especially late in the season when the seed pods are opening. ADULT: Large (BL 10–15 mm); slender; pale orange to bright red, with three large, black areas on the thorax and wings; underside has black spots; appendages black. NYMPH: Bright red; often occurring in dense clusters. RANGE: Warmer parts of the state; deserts, Central Valley, and coastal areas from San Francisco Bay southward.

Plate 3e. Common Milkweed Bug. *Lygaeus kalmii*. This is a smaller bug than the preceding species and is found on a wider variety of plants; sometimes large numbers aggregate in seed clusters of plants such as woody shrubs of the sunflower family. ADULT: BL 8–12 mm; head, thorax, and spots on abdomen velvety gray-black with abdomen and markings on the thorax and forewings red; wing membranes black, each with a conspicuous, median white spot and white lateral margins. NYMPH: Red with black markings. RANGE: Throughout much of the state at a wide range of elevations; adults occur outside the range of milkweed.

93. Big-eyed bugs. *Geocoris*. Several species of this genus are common on the ground under a wide variety of low-growing weedy plants as well as in alfalfa fields and other truck crops. Probably most feeding occurs on fallen seed, but big-eyed bugs sometimes eat soft-bodied insects such as leafhopper nymphs and mealybugs. ADULTS: Small (BL 3–4 mm); large, protruding eyes; head, thorax, and abdomen are of about the same width; pale tan or whitish with black eyes, or wholly black. RANGE: Throughout most of California.

94. Chinch bugs. *Blissus*. Several species occur in California, including the (**94**) **Southern Chinch Bug** *(B. insularis),* which is found on St. Augustine grass in the southern counties. It is closely similar to the **Chinch Bug** *(B. leucopterus),* one of the most notorious of all Hemiptera because it causes serious injury to wheat crops of North America. ADULTS: Small (BL 3–4 mm); slender, black, with the base of the antennae and the legs reddish or yellowish brown; wings either well developed, white with two median lateral spots, or short, covering only the base of the abdomen, brown with white tips. NYMPHS: Brown with a reddish abdomen bearing dark spots but variable in color between instars. All stages feed on wheat, corn, and other grasses, causing withering and drying of the plants by sucking the juices. RANGE: Native species occur in coastal grasses, sometimes causing damage to lawns; and in warmer parts of the state, San Francisco Bay area, Central Valley, and southern California.

Family Pyrrhocoridae

95. Bordered Plant Bug. *Largus cinctus*. This is a conspicuous and common bug in a wide range of habitats. ADULT: Large (BL 11–15 mm); mostly black with lateral margins of the thorax and forewings bright orange, and with a variable amount of orange speckling over the wings. NYMPH: Broadly oval, convex; bright metallic blue with a conspicuous, bright red spot at the base of the abdomen. RANGE: West of the

Sierra Nevada at low elevations and southern California. Especially common in beach and coastal strand habitats in association with lupine.

FLAT BUGS
Family Aradidae

Aradids are broad, flattened insects that are adapted to living under bark of dead trees. Apparently most species feed on wood-rot fungi, and both nymphs and adults are often found clustered together around bracket fungi on dead or living stumps and trees. About 30 species are known in California, and diligent collecting probably will reveal others.

96. Common flat bugs. *Aradus.* Most California aradids belong to this cosmopolitan genus. ADULTS: Small to moderately large (BL 4–11 mm); brown to blackish, mottled with tan or reddish brown; abdomen flared, more broadly so in female; thorax and abdomen sculptured with raised ridges; antennae short, thickened in some species; wings narrow, folded over middle of back. NYMPHS: Similar in outline, flatter with less sculpturing. RANGE: Forested areas; often found on bark of fallen conifers, frequently in association with wood-rot fungi, especially the nymphs. Members of another genus, *Mezira,* are more abundant in the foothills, often forming aggregations of nymphs and adults under loose bark of oak. ADULTS: Medium-sized (BL 6–9 mm); black or dark brown; somewhat rectangular in outline; the abdomen not flared, not as ornately sculptured as *Aradus.* RANGE: Throughout forested parts of the state; found under bark of conifers and hardwoods, and sometimes adults fly to freshly stained wood walls.

LACE BUGS
Family Tingidae

Members of the family Tingidae are small, flat bugs with the head and body hidden beneath a lacelike, rectangular shield, which consists of enlarged forewings and expanded thorax. The nymphs look quite different, usually much darker and covered with spines. All stages live together in colonies, often on undersides of leaves. Feeding causes drying of the leaves, and a colony's presence is evidenced by characteristic pepperlike frass pellets stuck to the paler, affected areas.

97. Common lace bugs. *Corythucha.* About a dozen species of this genus occur in California, each associated with a certain kind of native plant, including ceanothus, sycamore, Yerba Santa, sunflowers, walnut, alder, willow, etc. ADULTS: Small (BL 3–4 mm); white, yellowish, or pale greenish, sometimes mottled with brown. RANGE: Virtually throughout the state.

AMBUSH BUGS
Family Phymatidae

Ambush bugs are brightly colored, most often mottled in yellow and brown to blend with flowers in which they wait for their prey. They are deep-bodied, boat-hull shaped, with heavy, mantislike forelegs for grabbing the victims. Often insects much larger than themselves, such as Honey Bees and butterflies, are ambushed and quickly subdued by a powerful puncture of the beak.

98. Pacific Ambush Bug. *Phymata pacifica.* ADULT: Variable in size (BL 7–12 mm) and color, bright yellow to whitish with reddish, brown, or blackish markings. Common in flowers such as buckwheat *(Eriogonum)* and various members of the sunflower family but often overlooked because of their cryptic coloration and behavior. RANGE: Warmer areas, including mountains, Central Valley, and southern California coastal and desert areas.

PIRATE BUGS
Family Anthocoridae

99. Minute pirate bugs. *Anthocoris, Orius.* Several species of these genera are common in California. They live on foliage and flowers where they are predaceous on aphids and other small insects. ADULTS: Small (BL 2–4 mm); with an elongate, triangular head; black with a characteristic, white chevron pattern on the forewings. NYMPHS: Pink or salmon colored; often render annoying, pin-prick bites when one walks in heavily inhabited vegetation, such as alfalfa fields. RANGE: Throughout much of the state at low to moderate elevations.

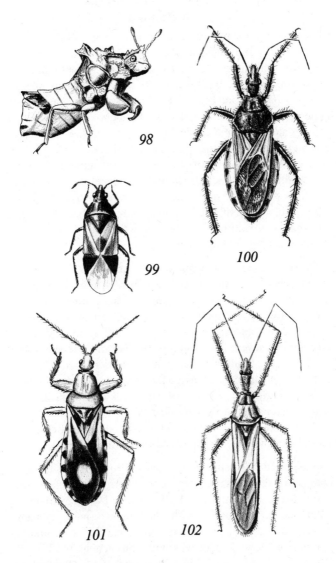

Figs. 98–102 True Bugs: 98, Pacific Ambush Bug, *Phymata pacifica;* 99, Minute Pirate Bug, *Anthocoris* species; 100, Spotted Assassin, *Rhynocoris ventralis;* 101, Western Corsair, *Rasahus thoracicus;* 102, Four-Spurred Assassin Bug, *Zelus tetracanthus*.

ASSASSIN BUGS, KISSING BUGS
Family Reduviidae

Members of the family Reduviidae are oval or elongate bugs with long legs and antennae. The forelegs are used for catching prey, and sometimes are modified mantislike (raptorial). Reduviids have large, beadlike, protruding eyes. Many species are spiny or hairy and bear sticky substances on the hairs which collect bits of plant material and other debris and aid in prey capture and in cryptic concealment of the bugs. Most species inhabit plants, are diurnal, and are slow fliers. They stalk their prey by slowly moving over vegetation or waiting at flowers. The victims, which include all kinds of insects, are snatched by quick movements of the forelegs and immediately subdued by a powerful venom injected through the beak. About 40 species are known in California, a few of which are nocturnal and normally suck the blood of mammals.

Plate 3f. Western Bloodsucking Conenose. *Triatoma protracta.* This large, nocturnal bug normally inhabits wood-rat nests but sometimes invades campsites and cabins. The adult flies readily and is attracted to lights, heightening possibility of entry into residences. Its bite causes a swelling and a systemic reaction and severe illness in sensitive persons. Related forms in Mexico and South America transmit the debilitating Chagas' disease, but virulent strains of this disease are not known in California. ADULT: BL 16–19 mm; black or brownish black; with an elongate head. RANGE: Warmer parts of the state including the Central Valley and oak-woodland and chaparral foothills of the Coast Ranges and southern California.

100. Spotted Assassin. *Rhynocoris ventralis.* This is the most commonly seen assassin bug in northern California during spring and early summer. ADULT: BL 9–14 mm; variable in color, usually red or tan with black thorax and wing membranes, and evenly spaced spots along the flared, protruding margins of the abdomen; legs with conspicuous black rings. RANGE: Siskiyou and Modoc counties south through the Sierra Nevada and the Coast Ranges to the desert margins of the mountains in southern California.

101. Western Corsair. *Rasahus thoracicus.* The painful bite of this bug is scarcely exceeded in severity by that of any other insect. The bug normally preys on other insects and bites humans only in self-defense. The Western Corsair is nocturnal and often attracted to lights on warm evenings. Because most reduviids rarely bite when handled, many collectors make the unforgettable mistake of picking up a Western Corsair while collecting at lights. ADULT: BL 18–23 mm; moderately slender; with smooth, shining body and legs, the front pair enlarged; tan or amber colored with black markings and wing membranes, each of which has a large, round tan spot. RANGE: Warmer parts of California, including Sierran foothills, coastal valleys, the Central Valley, and desert areas.

Plate 3g. Robust assassin bugs. *Apiomerus.* Several species of this genus occur in California. They are most often seen at flowers where they wait for insect visitors. ADULTS: Moderately large (BL 10–17 mm); thick-bodied with stout, hairy, tarantulalike legs. These bugs resemble spiders at first glance. The common species of the Coast Ranges is mainly black with red markings; red and yellow species occur in southern California and in the deserts. RANGE: Warmer parts of the state, from the San Francisco Bay area southward.

102. Four-spurred Assassin Bug. *Zelus tetracanthus.* Adults and nymphs of this genus accumulate a sticky secretion that gathers debris and aids in capturing prey. ADULT: BL 10–16 mm; slender, elongate with very thin legs and antennae; brown, gray, or black, paler individuals tending to have banded legs and abdomen; 4 pointed spurs occur in row across the back of the thorax. RANGE: Throughout warmer parts of the state, including both sides of the Sierra Nevada at moderate elevations, the Cental Valley, and deserts. The **Leafhopper Assassin Bug** *(Z. renardii)* is similarly widespread in California. ADULT: Usually smaller (BL 10–13 mm); green or gray when alive, with reddish wings and abdomen; thorax lacking spurs on hind dorsal margin; common in alfalfa fields and weedy areas where they prey on leafhoppers and other small insects.

BED BUGS
Family Cimicidae

Bed bugs used to be well-known domestic insects in all parts of the world inhabited by humans. However, with improved living conditions in western society during the past half century, these bugs have disappeared from all but the poorest residential areas. Most Californians have never seen a bed bug and will never be bitten by one. These bugs are small, oval, flat, dark brown, wingless or with wing stubs. Both nymphs and adults feed nocturnally on the blood of mammals or birds by means of a stout beak which pierces the skin and injects a secretion containing both an anticoagulant and an anaesthetic.

103. Bed Bug. *Cimex lectularius.* ADULT: BL 4–5 mm; brown, rust red, or purplish; covered with short bristles. They hide under mattress covers or in bedding by day, and their presence may be determined by dark stains of excrement on the sheets and a characteristic odor. Bites often occur in a slightly curving row of 3 evenly spaced punctures. RANGE: Unclean hotels, rooming houses, and residences; less common in California than in southern and eastern parts of the United States.

104. Swallow Bug. *Oeciacus vicarius.* ADULT: BL 3.5–4.5 mm; similar to the Bed Bug but clothed with longer, pale hairs. RANGE: Cliff Swallow and rarely Barn Swallow nests throughout the state. More than 1000 individuals may occur in one nest.

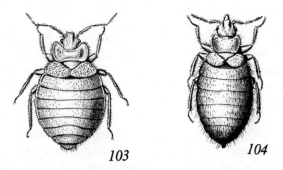

103   *104*

Figs. 103–104   Bed Bugs: 103, Bed Bug, *Cimex lectularius;* 104, Swallow Bug, *Oeciacus vicarius.*

PLANT BUGS
Family Miridae

Members of the family Miridae are the commonest Hemiptera in most areas of California. Many species attain high population densities, and most are specific to certain plants. Sweeping or beating a given kind of plant, such as grasses or oak, often will reveal large numbers of mirids. They are small, soft-bodied, elongate or oval bugs with prominent eyes and long, thin antennae and legs. Plant bugs actively run or fly, although some species have short-winged adult forms which resemble ants. Most mirids suck juices of soft plant parts via a long beak, but some are predators on other soft-bodied insects. More than 150 species are recorded in California, and others await discovery.

105. Ornate Plant Bug. *Closterocoris ornatus.* One of the largest (BL 7–9 mm) and most colorful mirids in California, this species is common on various native flowers. ADULT: Black with orange thorax and bright yellow lines on the wings. RANGE: Coast Ranges and coastal areas from southern Mendocino County southward; more prevalent in southern California.

106. Black grass bugs. *Irbisia.* Several species of this and related genera are common in grassy areas, particularly during spring. As the grasses dry, the bugs often migrate into nearby cultivated fields and gardens. ADULTS: BL 5–8 mm; shining black or grayish black with either black or red legs. RANGE: Coastal areas, inland valleys, and mountains throughout the state.

107. Tarnished plant bugs. *Lygus.* Members of this genus are among the commonest bugs in California, occurring in great numbers on a wide variety of flowers, fruit trees, truck crops such as cotton and alfalfa, and on weeds. ADULTS: BL 4–6 mm; winged; pale green or yellowish to dark brown or reddish, with a pale scutellum. NYMPHS: Pale yellow or green. RANGE: Throughout all but the deserts at a wide range of elevations. *L. elisus* and *L. oblineatus* are widespread species occurring abundantly in agricultural areas, particularly in the Central Valley.

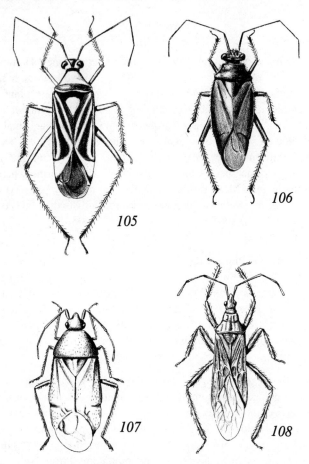

Figs. 105–108 True Bugs: 105, Ornate Plant Bug, *Closterocoris ornatus;* 106, Black Grass Bug, *Irbisia* species; 107, Tarnished Plant Bug, *Lygus* species; 108, Common Damsel Bug, *Nabis ferus.*

DAMSEL BUGS
Family Nabidae

108. Common Damsel Bug. *Nabis ferus.* This is a common inhabitant of weedy and cultivated fields in California. Damsel bugs are predaceous, feeding on smaller bugs such as aphids and leafhoppers. ADULT: BL 7–9 mm; slender; pale brownish gray; with somewhat mantislike but not enlarged forelegs. RANGE: Throughout low to moderate elevations in the state.

RIFFLE BUGS
Family Veliidae

Riffle bugs are tiny, oval insects having shorter legs than water striders. Veliids are found around banks and vegetation at the edges of streams and ponds, venturing out onto the water surface only when dislodged. Most species are restricted to freshwater places, but a few live around brackish or salt water. Riffle bugs eat small insects and other organisms, including those which live nearby and fall onto the water surface.

109. Minute riffle bugs. *Microvelia*. ADULTS: Small (BL 1.0–2.5 mm); flat, dull black bugs; winged forms less common than wingless adults, which resemble immatures. All are found together skittering in small herds across the water surface when their stream-side retreats are disturbed. Eight species occur in California, at least 2 of which are very common, but they are often overlooked because of their small size and secretive habits. RANGE: Mostly in the Coast Ranges and foothills of the Sierra Nevada up to about 1500 m (5000 ft) elevation.

SHORE BUGS
Family Saldidae

Saldids are small, oval, somewhat flattened bugs with moderately long legs used for running and jumping. Most are black or brown, although some have conspicuous white or yellow markings. These bugs inhabit lake shores, beaches, stream banks, and moss-covered rocks and tree trunks in wet places. About 30 species are known in California, and some are quite abundant, yet they are overlooked by all but the most careful observers. Many saldids are quick, running or flying at the slightest disturbance, while others are secretive, hiding in overhanging vegetation or other shady spots. Shore bugs are predators or scavengers, eating a variety of living or dead organisms found on the damp soil.

110. Black shore bugs. *Saldula*. The 22 species of this genus in California occupy all kinds of shoreline situations, including briny and alkaline places. ADULTS: Small (BL 3.5–5.0 mm to wing tips); body and basal half of wings blackish, distal part of

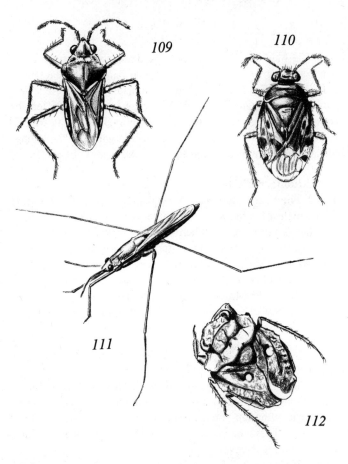

Figs. 109–112 True Bugs: 109, Minute Riffle Bug, *Microvelia* species; 110, Black Shore Bug, *Saldula* species; 111, Water Strider, *Gerris remigis;* 112, Toad Bug, *Gelastocoris oculatus.*

wings whitish or translucent. RANGE: Throughout the state including springs in desert areas. *S. pallipes,* which occurs all over North America and Europe, is the commonest species in California. It ranges from low-elevation salt marshes to over 3,000 m (10,000 ft), varying in size and color from one habitat to another.

WATER STRIDERS
Family Gerridae

Gerrids are elongate bugs that have long legs with the claws set well back from the tips, enabling the bugs to skate on water on little depressions in the surface film. Naturalists often observe the curious, symmetrical shadows cast by these depressions on the bottoms of ponds. Water striders move by synchronous, oarlike movements of the middle legs, which are longer than the others. Gerrids are predaceous, mainly on insects and other organisms that fall on the water, but they also catch aquatic insects that come to the surface. Adults of some species are wingless, but the degree of wing development varies greatly. Apparently this is related to dispersal and habitat stability; day length and temperature have been shown to influence development of winged forms. Generally, winglessness occurs only in species that live on permanent rivers, where flight dispersal is unnecessary. About 10 species are known in California, of which one occurs only on the Pacific Ocean, and two are restricted to the Colorado River.

111. Common Water Strider. *Gerris remigis.* This bug is a familiar sight on all kinds of freshwater surfaces, from eddies in fast-running streams to isolated stock tanks. ADULT: Large (BL 14–18 mm, length across leg span 30 mm); body brownish black, covered with velvety, waterproof pile that sometimes appears gray or silvery; winged or wingless. RANGE: The most widespread semi-aquatic insect in the United States and in California; occurs throughout the state, except in the deserts, up to about 2500 m (8,000 ft) elevation.

TOAD BUGS
Family Gelastocoridae

Gelastocorids are short, flattened, broad bugs with protruding eyes and a warty appearance, and they jump frog-hop fashion when approached. Their short, grasping front legs are used to catch prey, which includes insects and other small organisms. Two genera with 4 species occur in California.

112. Toad Bug. *Gelastocoris oculatus.* ADULT: BL 7–9 mm; mottled in gray, tan, greenish or red-brown, these bugs blend in with the sand along edges of streams and ponds. The variation in color pattern at one locality is often striking. RANGE: In virtually every county at a wide range of elevations.

BACKSWIMMERS
Family Notonectidae

Notonectids differ from all other aquatic insects in California by swimming upside down. The body is boat-hull shaped, elongate, with a flat underside and convex back. The eyes are large, occupying most of the head; the beak is short and capable of inflicting a powerful bite. A collector needs just one painful lesson in carelessly picking up a backswimmer. These bugs have both the front and middle legs adapted for grasping their prey, while the hindlegs are long and oarlike, flattened and fringed for swimming. Backswimmers are most commonly seen in pools of streams and springs, but they are good fliers and often appear in places such as swimming pools remote from other water. Prey consists of other insects and occasionally small fish, and some species are helpful in mosquito control.

113. Large backswimmers. *Notonecta.* Seven similar species of this familiar genus occur in California. ADULTS: BL 10–16 mm; robust; variable in color, whitish, or black and white to red-brown. RANGE: Throughout most of California, up to 2750 m (9,000 ft) elevation. Adults may be seen throughout the year. One species, the **Single-banded Backswimmer** *(N. unifasciata),* is widespread and occurs in a broad range of habitats, including saline pools and hot springs with temperatures as high as 36°C. **Kirby's Backswimmer** *(N. kirbyi)* is the largest and most widespread species in California. The **small backswimmers,** members of the genus *Buenoa,* are more slender and smaller (BL 5–7 mm) than *Notonecta.* ADULTS: Some species are either short winged or fully winged and may have two color forms at the same locality. Four species are known in California, of which the **Scimitar**

Backswimmer *(B. scimitra)* is the most common. It is a pale to smoky gray species with a large, sword-shaped stridulatory area on the foreleg. RANGE: Coast Ranges, Central Valley, and Sierra foothills, from Sonoma County to San Diego, and in the Imperial Valley.

CREEPING WATER BUGS
Family Naucoridae

Naucorids are oval, flattened bugs with mantis-type forelegs used for grasping their prey. They somewhat resemble the giant water bugs (Belostomatidae) but are smaller and lack the strap-like tail appendages characteristic of that family. Most naucorids inhabit slow streams with pebbly bottoms. They crawl or swim among the rocks in search of small organisms which constitute their food, such as water boatmen, mosquito larvae, and mollusks. These bugs can inflict an extremely painful bite. Six species in 2 genera occur in California.

114. California Creeping Water Bug. *Ambrysus californica.* ADULT: Small (BL 7–8 mm, BW 4.5–6 mm); winged; yellowish brown to greenish with darker markings. RANGE: Streams of the Coast Ranges from Mendocino County southward to the Transverse Range and San Bernardino Mountains. A similar species, the **Mormon Creeping Water Bug** *(A. mormon)*, is larger (BL 9–12 mm) and occurs throughout the northern two-thirds of the state at a wide range of elevations, in a variety of stream and lake habitats. The **Western Creeping Water Bug** *(A. occidentalis),* which is even larger (BL 10–12.5 mm), is the common species in still pools in southern California.

WATER BOATMEN
Family Corixidae

The family Corixidae has more species than any of the other aquatic Hemiptera. Populations often occur in huge numbers so that corixids are an important link in the food chain of water communities, converting plants and tiny animals into food for fish. Water boatmen possess a triangular, flat head and characteristically have alternating dark and light, transverse banding on the thorax. These insects resemble backswimmers to the casual glance, but corixids are flattened and swim right side up. Each pair of legs is adapted to a special function: the fore legs are used for food gathering, the mid-legs for clinging to substrates while feeding or resting, and the hindlegs are long, flattened, oarlike, and are used to propel the bugs through the water.

Water boatmen live in a wide range of habitats, from saline waters of Death Valley below sea level to frigid waters beneath ice in subarctic and high montane sites. Some species live exclusively in salt water, in the Salton Sea or along the coast in ocean bays; however, most corixids prefer fresh water. Usually these bugs forage in bottom ooze, eating algae, protozoa, and other microscopic organisms, but sometimes water boatmen are predators on larger animals, such as mosquito larvae. Corixids are preferred food of many fish, and the bugs and their eggs are harvested for human food in Mexico. Tons of nymphs and adults are used in bird, fish, and turtle petfood.

There are 7 genera and 25 species of Corixidae known in California, some of which are widespread and abundant.

115. Smooth water boatmen. *Trichocorixa*. ADULTS: Smaller than in other genera (BL 3–6 mm); with smooth, shiny wing covers and pronotum. Our 4 species are restricted to coastal marshes and either salty or fresh waters of the desert areas. A number of *Trichocorixa* prefer saline or alkaline water, including **(115) Salt Marsh Water Boatman** *(T. reticulata)*. ADULT: The smallest California corixid (BL 2.8–5.4 mm); females distinguished from all others by lacking the frosted white area along the lateral margin of the forewing. RANGE: Salt marshes and tidal pools along the coast, at the Salton Sea, and in alkaline waters of Death Valley where they inhabit the pool at Badwater, and the Colorado River Valley. Other common corixids in California include: The **Large Water Boatman** *(Hesperocorixa laevigata)*. ADULT: Large (BL 9–13 mm); dark, with strong, netlike lines on the forewings, especially in the mid-region. RANGE: One of the commonest corixids in California, *H. laevigata* occurs every-

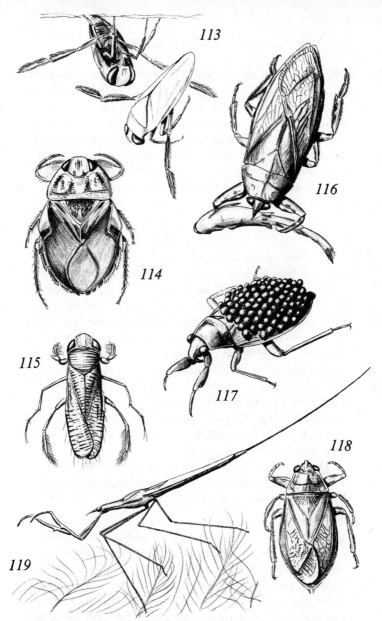

Figs. 113–119 Water Bugs: 113, Backswimmers, *Notonecta* species; 114, Creeping Water Bug, *Ambrysus californica;* 115, Salt Marsh Water Boatman, *Trichocorixa reticulata;* 116, Electric Light Bug, *Lethocerus americanus;* 117, Toe Biter, *Abedus indentatus* (male with eggs on back); 118, Little Toe Biter, *Belostoma flumineum;* 119, Common Water Scorpion, *Ranatra brevicollis.*

where except in the deserts. The **common water boatmen** *(Corisella)*. ADULTS: Small (BL 4–8 mm); pronotum roughened, wing covers nearly smooth with the dark reticulations forming irregular, longitudinal rows. RANGE: Throughout much of the state at a wide range of elevations; one species, *C. decolor,* is common in fresh and saline water ponds and is especially abundant in the rice fields of the Sacramento Valley, where countless numbers of adults are attracted to lights.

GIANT WATER BUGS
Family Belostomatidae

Members of the family Belostomatidae are brown, oval, flattened bugs with a pair of retractable, straplike breathing appendages at the end of the abdomen. Three genera occur in California, *Belostoma* and *Lethocerus* in standing water habitats and *Abedus* in streams. Belostomatids are strong swimmers, but normally perch on submerged objects to wait for prey. The grasping front legs are held in readiness and quickly seize the victims, including other insects, tadpoles, and even small fish and snakes. A paralyzing toxin is injected by the bug's strong beak. Some species occasionally are pests in fish hatcheries. They can bite a human, producing a burning sensation that lasts several hours and causes considerable reddening and swelling.

One of the most distinctive features of *Abedus* and *Belostoma* is the oviposition behavior; females deposit their eggs on the backs of males, who carry them until they hatch (**117**).

116. Electric Light Bug. *Lethocerus americanus.* ADULT: Huge (BL 45–60 mm to tips of wings); brown. Sometimes attracted to lights in large numbers. RANGE: Northern half of the state; there are older records from Riverside and Orange counties, but drying of freshwater ponds probably has eliminated this bug in southern California. A related species, *L. angustipes,* occurs in Death Valley but is primarily Mexican in distribution. Another large member of the genus *Lethocerus* is considered a great delicacy by peoples of the Orient, and one can purchase giant water bugs from food stores in Chinatown in San Francisco.

117. Toe Biter. *Abedus indentatus.* ADULT: Proportionately broader (BL 27–37 mm, width 14–20 mm) than other giant water bugs in California. Very common in riffle areas of streams, clinging to submerged vegetation or under debris. RANGE: Streams of southern two-thirds of the state, including the deserts.

118. Little Toe Biter. *Belostoma flumineum.* Bugs of this genus are smaller than *Abedus* and have a well-developed forewing membrane. ADULT: BL 18–23 mm, BW 9–12 mm. Found in ponds, springs, and slow creeks. RANGE: Principally in the Central, Coachella, and Imperial valleys. A similar species, *B. bakeri,* is more common and wide-ranging. It has been collected every month of the year in southern California.

WATER SCORPIONS
Family Nepidae

Water scorpions are green or brown, slender, cylindrical or somewhat flattened, and with long, thin legs, which are not modified for swimming. The common name derives from a long, tail-like air siphon on the end of the abdomen. These bugs are found in still-water habitats in tangled plant growth or debris where they are camouflaged by their sticklike appearance. They feed on a wide variety of aquatic animals, such as mosquito larvae and tadpoles, which they catch by waiting motionless until a victim swims within reach of the lightning-quick, grasping front legs. Only 3 species are known in California.

119. Common Water Scorpion. *Ranatra brevicollis.* ADULT: BL 30–40 mm, plus tail siphon 20–25 mm; elongate, cylindrical; brown; closing face of the foreleg femur with a broad, shallow excavation and a long process on the metathoracic underside nearly reaching the base of the abdomen. RANGE: Low-elevation streams from southern Humboldt County to San Diego, inland in the Sierran foothills and mountains of southern California. A related species, the **Desert Water Scorpion** *(R. quadridentata),* occurs in the Coachella and Imperial valleys. Its foreleg femur has a narrow, subapical notch defined basally by a strong tooth.

CICADAS, HOPPERS, APHIDS, SCALES, AND OTHERS
Suborder Homoptera

The suborder Homoptera contains a diverse group of insects with many structural and life history modifications. Leafhoppers and froghoppers are relatively unmodified, while planthoppers and treehoppers often possess grotesquely enlarged and complex head or pronotal structures. Females of scale insects and mealybugs are greatly modified, completely wingless, and often fixed in place. Life history specializations include the seasonal alternation of bisexual and parthenogenetic, all female, generations of aphids. In some forms the mouthparts are poorly developed and nonfunctional in the adults; others secrete waxy substances that may cover the body or form projecting filaments or frothy masses. Many species secrete liquids rich in sugar ("honeydew").

References

Ferris, G. F. 1937–1955. *Atlas of the Scale Insects of North America.* Palo Alto: Stanford University Press; 7 vols.

McKenzie, H. L. 1956. The armored scale insects of California. *Calif. Insect Survey, Bull.* 5:1–209.

———. 1967. *Mealybugs of California.* Berkeley: University of California Press; 525 pp.

Oman, P. W. 1949. The nearctic leafhoppers (Homoptera: Cicadellidae), a generic classification and checklist. *Ent. Soc. Washington Memoir* 3; 253 pp.

Simons, J. N. 1954. The cicadas of California. *Calif. Insect Survey, Bull.* 2:151–196.

Young, D. A. 1977. Taxonomic study of the Cicadellinae (Homoptera: Cicadellidae) II. New World Cicadellinae and the genus *Cicadella. No. Car. Agric. Exp. Sta., Tech. Bull.* 239; 1135 pp.

CICADAS
Family Cicadidae

The most characteristic feature of cicadas is sound production. The males possess organs (tymbals) in the base of the abdomen which emit a high-pitched, shrill cry. The sound waves are generated by strong muscles vibrating a taut membrane at a high frequency, something like a drumhead. Specific "song" patterns play a role in courtship. The small males of the genus *Platypedia* lack tymbals and can emit only short clicking noises by rattling their wings together.

Cicadas are notoriously long-lived, the nymphs requiring 2–5 years to develop (up to 17 years in one eastern species). These immatures burrow underground and feed on roots.

Some 65 species of cicadas live in California, mostly in drier habitats, the deserts, brush-covered hills, and forest clearings, although some species extend into agricultural areas as well. Population explosions often occur in some species (e.g., *Okanagana cruentifera* on Great Basin Sage Brush in Inyo and Mono counties). This tendency and their loud songs make cicadas one of our more conspicuous insects.

Plate 3h. Woodland cicadas. *Platypedia.* Members of this genus are relatively slender cicadas characteristic of foothill and mountain areas. ADULTS: BL 20–30 mm; generally black or bronzy colored with yellow, orange, or reddish legs and body markings; often with a distinctly colored line across the hind margin of the prothorax. Females oviposit into twigs of native trees (oaks, Madrone, willows, etc.), shrubs such as *Baccharis,* and sometimes in deciduous orchard trees. RANGE: Northern mountains, Coast and Peninsular ranges. Most of the 18 species recorded in California occur in the northern half of the state.

120, 121. Apache Cicada. *Diceroprocta apache.* The deafening chorus of males of this species, emanating from roadside Tamarisk and Palo Verde, is a familiar sound to motorists passing through the lower deserts in the summer. ADULT (**120**): BL 25 mm, WL 35 mm; generally brown or blackish with contrasting yellow transverse collar; veins conspicuously lighter over basal half of wing. Hosts are mainly native desert shrubs, but oviposition punctures by the females cause minor damage to elms, date palms, and citrus in the Coachella Valley. NYMPH (**121**): Body heavy, abdomen down-curved; forelegs thickened and clawed for digging; dark brown. Lives underground, sucking sap from roots of plant host; life cycle requires 2 years. RANGE: Limited to Colorado Desert, Imperial and Riverside counties.

Plate 4a. Red-winged Grass Cicada. *Tibicinoides cupreosparsus.* ADULT: Small (BL 20 mm; WL 15 mm); body black, with bright red areas at bases of the blackish clouded wings. Male is a feeble singer; found in grass-covered hillsides; probably uses various species of grasses and herbaceous plants as hosts both in the adult and nymphal stages. RANGE: Brush

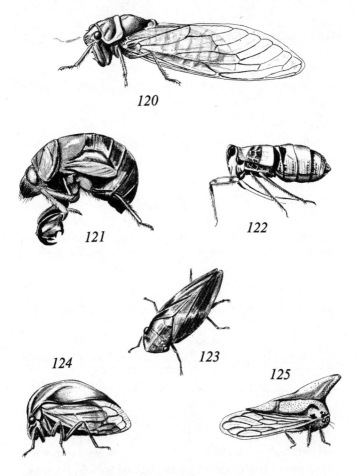

Figs. 120–125 Homopterans: 120, Apache Cicada, *Diceroprocta apache* (adult); 121, *D. apache* (nymph); 122, Dammers' Planthopper, *Loxophora dammersi;* 123, Annulate Spittle Bug, *Aphrophora annulata;* 124, Three-Cornered Alfalfa Hopper, *Spissistilus festinus;* 125, Oak Treehopper, *Platycotis vittata.*

and grasslands of southern California, Los Angeles, San Diego, and western Riverside counties.

PLANTHOPPERS
Superfamily Fulgoroidea

122. Many planthoppers are difficult to distinguish from leafhoppers and spittle bugs. They are normally flatter and broader, have antennae arising on the sides of the head beneath the eyes (rather than in front of or between), and often have complex (usually snoutlike) projections, keels, or ridges on the front of the head. Primarily a tropical group, California has only about 100 species. One of the most curious is (**122**) **Dammers' Planthopper** *(Loxophora dammersi)* because of its amusing antics and shape. Holding its snout erect, it runs in all directions and tries to hide on the far side of its perch when teased. It is small (BL 4 mm), lacks a head horn, and is pale greenish with transverse brownish bands. Its host is *Agave* and its range is the Colorado Desert.

SPITTLE BUGS
Family Aphrophoridae (Cercopidae)

123; Plate 4b. Nymphs of the family Aphrophoridae are more often encountered than the adults but are seldom seen because of the thick, white, frothy mass surrounding them (**Plate 4b**). These masses are most common on herbaceous plants which provide food for the nymphs and adults. The latter resemble leafhoppers but characteristically sit froglike with the head elevated; structurally they differ in being broader across the middle rather than parallel-sided and by having only 1–2 stout spines on the hind tibia. California has only about 6 species, all generally similar to the (**123**) **Annulate Spittle Bug** *(Aphrophora annulata)* which is medium-sized (BL 11 mm) and brown with irregular, darker brown markings. It may be found on almost any plant throughout the state except at the highest elevations.

LEAFHOPPERS
Family Cicadellidae

Many plant pests are members of the family Cicadellidae, which has more than 200 species in California. Their feeding can remove enough sap to wilt and kill growing tips, and sometimes transmits disease-producing viruses. Leafhoppers are agile, hopping insects, often running sideways, and many are brightly colored (although many are cryptically colored). Unlike their close relatives, the spittle bugs, they are slender (thorax scarcely wider than wing margins), the nymphs never cover themselves with froth, and the hind tibiae have 1 or more rows of small spines.

Plate 4c. Blue Sharpshooter. *Hordnia circellata.* A medium-sized (BL 7 mm), bright blue or blue-green species, often seen on garden plants. Other notable species in the state include the **Beet Leafhopper** *(Circulifer tenellus),* which is small (BL 3.5 mm); pale green to brown with slightly translucent wings; a major pest of truck crops, especially sugar beets, spinach, and tomatoes, to which it transmits Curly Top, a very destructive virus disease. The **Clover Leafhopper** *(Aceratagallia sanguinolenta)* is small (BL 3 mm); flattened; light gray mottled; most serious on legume crops. The **Apple Leafhopper** *(Empoasca maligna)* is small (BL 3.5 mm); yellow green, with dashes and lines of darker shade on body and wings; affects many crops.

TREEHOPPERS
Family Membracidae

Adult treehoppers have the front portion of the thorax greatly expanded to cover the abdomen, even extending upward and laterally to form various thornlike projections. With such shapes and their brown or green colors, they easily escape detection on vegetation. The nymphs are characterized by numerous erect spines on the back of the abdomen. Both immatures and adults jump readily. California has about 40 species, several of which injure plants through their egg punctures.

124. Three-cornered Alfalfa Hopper. *Spissistilus festinus.* One of the most common and widespread species, found on a wide variety of agricultural and ornamental plants. ADULT: Small (BL 5 mm); bright green, with a rounded, triangular thorax bearing short, anterolateral spines. RANGE: Widespread, especially along watercourses, in the southern half of California, including the deserts.

Plate 4d. Keeled Treehopper. *Antianthe expansa.* ADULT: Small (BL 5 mm), green; distinctive because the thorax is greatly expanded to form a sharp keel (complete to the head) and long anterolateral spines. RANGE: Spotty in more moist portions of southern California. Adults are often seen in vegetable gardens in association with the black, spiny nymphs.

125; Plate 4e. Oak Treehopper. *Platycotis vittata.* Compact aggregations of nymphs or adults of this species are found on lower branches of deciduous and live oaks in spring and occasionally on other broadleafed trees. The eggs are inserted into small branches; females remain with their eggs until they hatch and maintain contact with the aggregations of nymphs (**Plate 4e**). ADULT (**125**): BL 9–12 mm to tips of wings; pale bluish gray with the surface of the pronotum finely punctured, with a long, median horn and with longitudinal reddish orange stripes, or dull bronzy with a short horn and reddish dots. NYMPH (**Plate 4e**): Black with yellow and red markings and 2 soft, black spines on the back. RANGE: Foothills of the Coast Ranges and Sierra Nevada.

APHIDS
Family Aphididae

Aphids are small, soft-bodied insects, well known to gardeners and farmers because of their destructive feeding habits; there are many economically important species. Consistent characteristics which distinguish aphids are a pear-shaped body with a pair of tubular, hornlike projections (cornicles) near the posterior end. The winged forms have two pairs of membranous wings, the fore pair much larger than the hind, both with few veins. When at rest the wings are held vertically over the body.

Aphid life cycles are complex, with winged and wingless forms as well as sexual and parthenogenetic reproductives appearing at different times or alternating regularly according to season. Individuals have the habit of discharging a sugary solution (honeydew) from the anus. Ants are attracted to honeydew and foster the spread of aphids. California has over 450 species.

126–128. Rose Aphid *(Macrosiphum rosae)* ADULT: Comparatively large (BL 3 mm) with green body and contrasting black cornicles. This is the nemesis of the rose gardener; dense colonies form on the buds and terminal shoots, usually in the springtime. RANGE: Virtually everywhere in the state where the hosts are grown.

129. Spotted Alfalfa Aphid. *Therioaphis maculata.* ADULT: Small (BL 1.5 mm); pale yellow with rows of dark spots on the back. It is a serious pest of alfalfa and occurs on other legumes. RANGE: Introduced to North America from the Middle East, reaching California in 1954 and spreading throughout alfalfa growing regions of the state by 1958.

130. Giant Willow Aphid. *Tuberolachnus salignus.* ADULT: Large (BL to 4.2 mm); brown with black spots, including a larger spot in the center of the abdomen; cornicles short, surrounded by larger black spots; body covered with a fine, waxy powder. The species feeds in colonies on the trunks and branches of willow. Adults and nymphs kick their hind legs vigorously when disturbed. RANGE: Throughout the state where willows are found. The **Deodar Aphid** *(Cinara curvipes)* is similar in appearance, very large (BL to 5 mm), dark gray with black mottling. It forms dense colonies on the undersides of Deodar Cedar *(Cedrus deodara)* branches and its native host, firs *(Abies)*. RANGE: Mountain distribution of *Abies* and in urban situations where its hosts are grown.

SCALE INSECTS, MEALYBUGS
Superfamily Coccoidea

Scale insects and mealybugs, like the aphids, cause extensive damage to California agriculture. Many of our 200-plus species become so numerous on cultivated plants, especially orchard trees, shrubs, and ornamentals, that they may kill the host.

The group exhibits strong sexual dimorphism: males are very small, delicate, white insects having one pair of wings

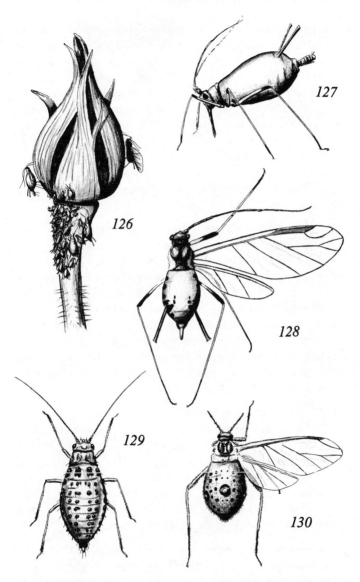

Figs. 126–130 Aphids: 126, Rose Aphid, *Macrosiphum rosae* (colony on rosebud); 127, *M. rosae* (apterous female); 128, *M. rosae* (alate reproductive); 129, Spotted Alfalfa Aphid, *Therioaphis maculata;* 130, Giant Willow Aphid, *Tuberolachnus salignus*.

and long, hairlike processes on the tip of the abdomen; females are obese, rounded forms, lacking wings, a separate head, or body segments and with minute, nonfunctional appendages at maturity. Adult females of the "armored scales" live under a scalelike covering secreted by their numerous wax glands. The unarmored or "soft scales" do not form a scale and are often rather large (BL 5–8 mm). The adult females of both types remain fixed in place by their nonretractable, filamentous mouthparts and often also gluelike secretions. Dispersion is accomplished by the young, actively crawling nymphs (**131**). (The mealybugs remain active throughout life and are an exception to the foregoing.)

131–134. California Red Scale. *Aonidiella aurantii.* This armored scale is the most important citrus pest in the world. FEMALE (**132**): Small (1.8 × 2 mm); very flat, kidney-shaped, with abdominal portion on indented side; legless; red body color shows through clear, circular scale (**133**). Attaches to all parts of plant, often conspicuous on fruit. MALE (**134**): Tiny (BL 1 mm), covered with powdery, white wax. Hosts include a wide variety of native and cultivated plants, but damage is most serious to citrus. RANGE: Southern half of state, primarily in the coastal citrus belts. Thought to be introduced from Australia before 1880.

135, 136. San Jose Scale. *Quadraspidiotus perniciosus.* FEMALE (**136**): Small (BL 2 mm); oval, disc-shaped, with two pairs of lobes; body bright yellow; scale (**135**) grayish, circular, irregular, with central nipple. The most serious scale pest of deciduous fruit trees in the state; lives on a large number of hosts, making control difficult. RANGE: Supposedly introduced at San Jose from China about 1870, now troublesome throughout the state.

137. European Elm Scale. *Gossyparia spuria.* FEMALE: An unarmored scale insect; large (BL 3–4 mm); body oval; shiny, dark reddish brown with thick, white, cottony fringes. Dense masses congregate on the undersides of elm tree limbs. In the spring, their copious release of honeydew falls like rain and fosters growth of black smut fungus which stains plants and

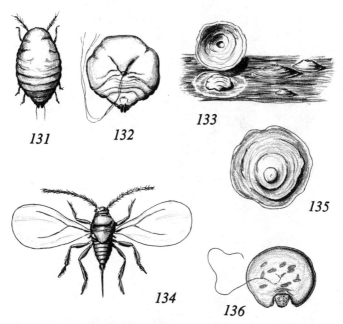

Figs. 131–136 Scale Insects: 131, California Red Scale, *Aonidiella aurantii* (crawler); 132, *A. aurantii* (female); 133, *A. aurantii* (scale lifted from female); 134, *A. aurantii* (male); 135, San Jose Scale, *Quadraspidiotus perniciosus* (scale); 136, *Q. perniciosus* (female).

sidewalks with sticky residue. Sap depletion of host may cause leaf drop and kill twigs. RANGE: A European species, now widespread in the state.

Plate 4f. Cottony-cushion Scale. *Icerya purchasi*. Feeds on a wide variety of plants. Before the turn of the century, it nearly destroyed the citrus industry in southern California, earning this species a place among agriculture's worst pests. A native of Australia, it was introduced at Menlo Park in 1868 or 1869. The growers were able to stop it only with help from the **(392)** Vedalia Beetle, a natural predator brought from Australia in the 1880s. This was an early example of the technique of biological control. FEMALE: An unarmored scale easily recognized by the elongate, fluted or grooved, white, waxy egg sac; large (BL 5–7 mm; with egg sac, 10–15 mm); egg sac

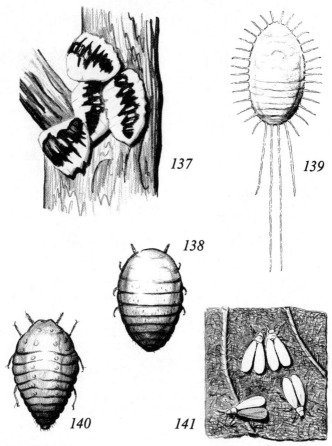

Figs. 137–141 Homopterans: 137, European Elm Scale, *Gossyparia spuria* (females); 138, Cochineal Scale, *Dactylopius* species (female); 139, Long-Tailed Mealybug, *Pseudococcus longispinus* (female); 140, Grape Phylloxera, *Phylloxera vitifolia* (female); 141, Greenhouse White fly, *Trialeurodes vaporariorum*.

contains hundreds of bright red, oblong eggs; body color reddish brown, legs black; tufts of short, black hairs in parallel rows along edge of body. RANGE: Still widely distributed in California and a potential danger to citrus.

138; Plate 4g. Cochineal scales. *Dactylopius.* One member of the genus, *D. coccus,* was long used by Mexican Indians for preparing a crimson dye. After its discovery by the Spanish, it was cultured in other parts of the world and its host, Prickly Pear Cactus, became widely dispersed. In Australia, the spread of the plant became disastrous to ranchlands. A cactus-feeding pyralid moth from South America, *Cactoblastis cactorum,* was successfully introduced to control it. Synthetic dyes finally replaced the product and killed the industry. ADULTS (**138**): Unarmored; medium-sized (BL 2–3 mm); body oval, convex, flattened beneath; abdominal segmentation distinct; red or crimson, covered by masses of white, cottony wax secretions. The latter is conspicuous on flat cactus *(Opuntia)* pads in the springtime (**Plate 4g**). RANGE: Several species are common in the deserts and other dry zones of southern California.

139. Mealybugs. *Pseudococcus* and related genera. ADULTS: Mealybugs are like unarmored scale insects but are usually larger (BL 2–4 mm), have well-developed legs, and are quite mobile. They also are covered with a frosty secretion of white wax and with fingerlike lateral and terminal filaments. They frequently congregate on the roots, stems, and leaves of citrus, garden, and greenhouse plants, causing harm by extracting sap and depositing honeydew. RANGE: About 170 species are known throughout the state, occurring on many kinds of plants. Perhaps the most general is the (**139**) **Long-tailed Mealybug** *(Pseudococcus longispinus).*

PHYLLOXERANS
Family Phylloxeridae

Phylloxerans are similar to aphids but are without cornicles and hold their wings horizontally at rest. They are equally destructive plant pests, especially the following notorious species.

140. Grape Phylloxera. *Phylloxera vitifolia.* A native of the eastern United States, this species inhabits the roots of indigenous grapes without doing serious harm. After being carried to the West (reaching California in 1852), the insect proved so destructive to imported European grapes that it nearly destroyed our wine industry. Use of resistant rootstock finally limited its damage. ADULT: Minute (BL 1 mm); oval; yellowish green or brown; 4 forms, with or without wings, and with different habits. Causes galls on grape roots and leads to the slow death of the vine. RANGE: Widespread but a more severe pest in the vineyards of the moister northern half of the state than in the south.

WHITEFLIES
Family Aleyrodidae

141; Plate 4h. Both wings and body of these minute (BL 2–3 mm) homopterans are covered with a fine, white, powdery wax. The wings at rest are held rooflike over the body and the insect resembles a tiny moth. Male and female are winged and similar. The nymphs (**Plate 4h**) of many species resemble scale insects and secrete flat wax plumes or ribbons from the back. Radiating from the body, these excrescences make the insect look like a tiny flower.

Whiteflies occur in dense colonies, usually on the undersides of leaves of citrus, greenhouse, and other valuable plants. Sap feeding by adults and nymphs causes leaf curl and general decline. Damage occurs also from heavy honeydew deposition which encourages the growth of sooty fungus.

This is primarily a tropical group and relatively few species live in our state. An exemplary species is the **Greenhouse Whitefly** *(Trialeurodes vaporariorum)*. ADULT (**141**): BL 1.5 mm; body pale yellow, wings pure white. Found on a large variety of host plants; very common and often most destructive to soft garden plants and some trees. RANGE: Widespread in agricultural portions of state.

THRIPS
Order Thysanoptera

Thrips are tiny insects, commonly less than 1 mm long, although a few range up to 5 mm. They are mostly yellow, brown, or black, and are found on vegetation, particularly in flowers. Adults sometimes are wingless but usually have 4 very narrow wings with reduced venation and long marginal fringes. Many thrips exhibit the behavior of curving the apex of the abdomen upwards, and in winged individuals this is preparatory to flight. In some respects the metamorphosis is of an intermediate type. The early stages, or nymphs, resemble the adults in general form, but there are one or two active or inactive pupal instars, and some species spin a cocoon in which the adult develops.

The vast majority of species feed by sucking juices from living plants, and many are quite specific as to the kind of plant inhabited. Some species are predaceous, and others live in moist, decaying plant habitats such as under bark of dead trees. Thrips often occur in tremendous numbers and can cause severe injury to crop plants by drying up blossoms or developing fruit. A few transmit plant virus diseases.

There are 2 suborders, the Terebrantia, in which females possess a sawlike ovipositor, and the Tubulifera, in which both sexes have the terminal abdominal segments tubelike. The latter group contains only about one-third of the more than 180 species of thrips recorded in California; but Tubulifera are more secretive in habits, fewer are of economic concern, and it is likely that many more species await discovery.

References

Bailey, S. F. 1957. The thrips of California. Part I: Suborder Terebrantia. *Calif. Insect Survey, Bull.* 4(5):143–220.

Cott, H. E. 1956. Systematics of the suborder Tubulifera (Thysanoptera) in California. *Univ. California Publ. Ent.* 13; 216 pp.

Lewis, T. 1973. *Thrips, Their Biology, Ecology and Economic Importance.* New York: Academic Press; 366 pp.

Suborder Terebrantia

Family Thripidae

142. Citrus Thrips. *Scirtothrips citri.* ADULT: Small (BL about 0.9 mm); yellow or orange with colorless wings. RANGE: Central Valley and Coast Range valleys from Lake County southward. This appears to be a native insect that adapted to introduced subtropical fruits. It occurs on many native plants such as Chamise and willow, but it is most noticeable on citrus, where it causes deformation of the fruit.

143. Western Flower Thrips. *Frankliniella occidentalis.* ADULT: Small (BL about 1.0 mm); orange-yellow with dusky

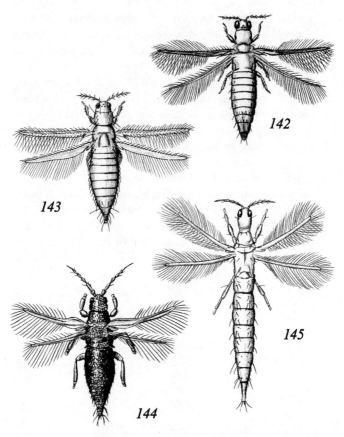

Figs. 142–145   Thrips: 142, Citrus Thrips, *Scirtothrips citri;* 143, Western Flower Thrips, *Frankliniella occidentalis;* 144, Greenhouse Thrips, *Heliothrips haemorrhoidalis;* 145, Black Hunter, *Leptothrips mali.*

markings on the sides of the abdomen, with large eyes, and covered with stout bristles. RANGE: The most widespread thrips in California; occurs almost throughout the state in a wide variety of habitats, from Mt. Whitney to the Imperial Valley, and in innumerable kinds of flowers.

144. Greenhouse Thrips. *Heliothrips haemorrhoidalis.* ADULT: BL about 1.25 mm; dark brown to almost black; with a reticulated surface and transparent wings. RANGE: In greenhouses throughout the world and outdoors in areas with mild climate, such as coastal parts of California. It feeds in and often damages a wide range of ornamental plants.

Suborder Tubulifera

Family Phlaeothripidae

145. Black Hunter. *Leptothrips mali*. This species occurs on a wide range of plants where it is predaceous on a variety of mites, scale insects, aphids, and moth eggs. ADULT: Relatively large (BL 1.8–2.6 mm); dark brown to black with clear, white wings that appear as a white line down the back when folded. RANGE: Throughout most of California, on both sides of the Sierra Nevada at low to moderate elevations, to the margins of the deserts.

NERVE-WINGED INSECTS
Order Neuroptera

The Neuropterans are characterized by 4 membranous wings with many branched, longitudinal veins connected by a great many crossveins. All have chewing mouthparts and are predaceous (except some adults which do not feed). There are 3 suborders, 12 families, and more than 150 species represented in California.

DOBSONFLIES AND ALDERFLIES
Suborder Megaloptera

Members of the suborder Megaloptera are medium-sized to large insects, which are found around streams that the larvae inhabit. The adults have long wings which are folded rooflike over the back when at rest, and they are most often seen clinging to streamside vegetation. Males of some species have greatly enlarged, scythelike mandibles, but they are harmless. The larvae ("hellgrammites") are exclusively aquatic and are distinguished by the tail-like last segment and slender, tapering lateral processes on the abdominal segments. Tufts of filaments on or at the bases of these processes serve as gills, and hellgrammites spend their entire lives under water. They have large, powerful mandibles and are fiercely predaceous on all kinds of small animals. They seek prey in the bottom mud or under debris. When mature, they burrow into banks above the water and pupate in earthen cells. The adults do not enter the water but deposit the eggs in masses on objects overhanging the aquatic habitats.

DOBSONFLIES
Family Corydalidae

Plate 5a. California dobsonflies. *Neohermes.* ADULTS: Large (BL 40–55 mm to wing tips; wing expanse 8.5–10.5 cm); wings gray with a transverse, variegated pattern of interrupted black lines. Antennae of males are as long as the body and have beadlike segments, each with a band of erect hairs around it. Nocturnal and often attracted to lights near streams during summer. LARVAE: Elongate (BL to 40 mm); cylindrical; body white, head and thorax brownish; abdomen with lateral, fingerlike processes on segments 1–8; last segment with a pair of lateral hooks. Hide in leaf and twig litter in small streams. RANGE: Two similar species occur widely in mountainous areas of the state: *N. filicornis* in southern California and the central Coast Range; *N. californica* in the Sierra Nevada, foothills to middle elevations. Several species of another genus, *Protochauliodes,* are easily confused with *Neohermes.* They are similar in general appearance but differ by having threadlike antennae in both sexes.

ALDERFLIES
Family Sialidae

146. California Alderfly. *Sialis californica.* ADULT: Medium-sized (BL 10–15 mm); wings and body smoky black. Fly sluggishly during the day amid streamside vegetation. LARVA: Similar to preceding but much smaller (BL 10–15 mm); abdomen with lateral filaments on segments 1–7 only and terminating with a long, median filament or tail. Aquatic, in stream riffles, debris, and pools. RANGE: Mountains and hills of northern and western California. Five quite similar species occur in the state.

SNAKEFLIES
Suborder Raphidioptera

Adult snakeflies are characterized by the prolonged head and very long first thoracic segment, which are movable and give them a serpentlike appearance. The wings are transparent with a conspicuous dark spot on the anterior margin near the tip.

Plate 5b. Common snakeflies. *Agulla*. ADULTS: Medium-sized (BL 12–25 mm to wing tips); ocelli present; wing spot containing a crossvein. Common on shrubs in the spring and summer. LARVAE: Elongate (BL 15–20 mm); head and first thoracic segment shiny black; abdomen pinkish brown; slightly swollen at middle. Found under loose tree bark and in other protected habitats where they prey on other small insects. RANGE: Generally distributed through state; more common in cooler climes and mountains. About a dozen similar species are known in California.

Suborder Planipennia

GREEN LACEWINGS
Family Chrysopidae

Plate 5c. ADULTS: A large group of species of small (BL 10–15 mm), delicate, green, yellow-green, or brown neuropterans with transparent, shining wings (anterior marginal crossveins of FW unbranched) and copper or golden-colored, shining eyes. They are common on foliage and often fly to light at night. LARVAE ("aphid lions"): Small (BL up to 10 mm); tapered toward both ends; mottled whitish and tan to red-brown; thorax dark, with two pairs of light bumps on each side; head with a pair of conspicuous, sickle-shaped jaws used to capture and drain body juices of prey, which includes aphids; legs with a trumpet-shaped extension between the paired, apical claws. Larvae of some species have the curious habit of entangling carcasses of victims and other debris among their hooked body hairs, thus acquiring a kind of camouflage. EGGS: Oval, white or greenish, elevated above the surface on long (5–10 mm) stalks. RANGE: General, common even in urban areas. One of the gardener's best friends because of the larva's voracious appetite for plant pests, especially Homoptera. Adults feed on honeydew, pollen, or are predaceous. A representative species is the (**Plate 5c**) **Common Green Lacewing** *(Chrysopa carnea).*

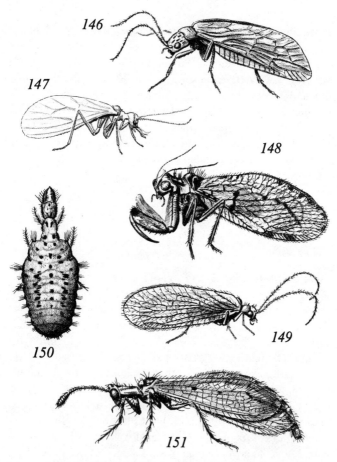

Figs 146–151 Neuropterans: 146, California Alderfly, *Sialis californica;* 147, Dustywing, *Coniopteryx* species; 148, Brown Mantispid, *Plega signata;* 149, California Spongilla-Fly, *Climacia californica;* 150, Common Antlion, *Brachynemurus* species (larva); 151, *Brachynemurus* (adult).

BROWN LACEWINGS
Family Hemerobiidae

Plate 5d. ADULTS: Brown lacewings resemble green lace-wings in general form and habits except for usually smaller size (BL 5–10 mm), sombre brown or yellowish colors, and anatomical details (anterior marginal crossveins in FW forked). LARVAE: Similar to chrysopids but more slender-bodied and with smaller heads and mandibles; distinguished by lack of trumpet-shaped extension between the paired tarsal claws, and lack of hooked body hairs. Feed on small, soft-bodied insects and mites, including many agricultural pests such as spider mites, aphids, and other Homoptera. EGGS: Placed singly directly on the substratum, not elevated on stalks. RANGE: General. A typical species is (**Plate 5d**) **Pacific Brown Lacewing** *(Hemerobius pacificus),* which has tolerance to low temperatures and is active throughout the year in coastal areas.

DUSTYWINGS
Family Coniopterygidae

147. Common dustywings. *Coniopteryx.* ADULTS: Dusty-wings might be confused with (**141**) whiteflies because of their small size (BL 2–3 mm) and powdery wax covering; but con-iopterygids have long, threadlike antennae and chewing mouthparts. The simple wing venation with few crossveins distinguishes them from other Neuroptera. Dustywings are mainly arboreal and often occur on conifers, oaks, and other trees, as well as on many other kinds of plants. LARVAE: Smooth; large thorax and small, strongly tapering abdomen. Prey actively on tiny, plant-feeding insects and mites and are beneficial to horticulture. RANGE: They are seldom noticed because they are small, but there are more than 20 species known from various parts of California, several of them re-cently discovered.

MANTISPIDS
Family Mantispidae

Mantispids are named for their enlarged grasping forelegs, like those of mantids. Among the several species in the state, one is of broad distribution; the others are confined to the desert areas.

148. Brown Mantispid. *Plega signata*. ADULT: Medium-sized (BL 22 mm to wing tips); wings speckled with black. Nocturnal, often attracted to light. LARVA: Unknown; pupae have been found within moth pupae in the soil. Related species develop in nests of wasps and bees. RANGE: Restricted to desert portions of southern California.

SPONGILLA-FLIES
Family Sisyridae

149. California Spongilla-fly. *Climacia californica*. ADULT: Similar to (**Plate 5d**) brown lacewings but small (BL 6 mm) and with the 2 major, anterior, longitudinal wing veins fused some distance from the wing tip. Found near water, and adults may be attracted to light at night. LARVA: Form basically similar to that of lacewings but mouthparts joined in front of the head, mandibles very long and needlelike; several pairs of segmented gills on the abdomen. Aquatic, in lakes. Lives on freshwater sponges, piercing the cells for food. RANGE: Known in California only from Clear Lake but may occur in other northern lakes where the host sponges occur. The Clear Lake population may have been exterminated by insecticide treatment for the Clear Lake Gnat.

MOTH LACEWINGS
Family Ithonidae

Plate 5e. Moth Lacewing. *Oliarces clara*. ADULT: Medium-sized (BL 22–34 mm to wing tips, wing expanse 25–40 mm); body dark brown to blackish; wing membrane milky, translucent, shiny; tufts of black hair protruding from sides of thorax. Emerge by the hundreds during late morning and early afternoon in April and May. LARVA: Unknown, except for the newly hatched stage. It is speculated that larva is subterranean and predatory on other soil insects. RANGE: Known only from desert canyons of Imperial and Riverside counties. This is a rare insect whose life history is almost a complete mystery.

LARGE LACEWINGS
Family Polystoechotidae

Plate 5f. Spotted Large Lacewing. *Polystoechotes punctatus*. ADULT: Large (BL 22–38 mm to wing tips); hairy, blackish with broad wings marked with numerous small, irregular, dark spots; longitudinal veins of central area of wing mostly parallel, only one staggered line of crossveins. Often attracted to campfires or lights at night in wooded areas. The larval habits are unknown. RANGE: Sierra Nevada, White Mountains, and Cascade and North Coast ranges, at middle elevations.

ANTLIONS
Family Myrmeleontidae

Antlions are well represented in all parts of the state, with more than 50 species. They are best known from the larval habits of certain genera (only *Myrmeleon* in California), which construct funnel-shaped pit traps in sandy soil (**Plate 5g**) to catch ants and other small insect prey. Most species occur in the more arid southern portions of California where the dominant genus is the following.

150, 151. Common antlions. *Brachynemurus*. ADULTS (**151**): Medium-sized (BL 35–50 mm); abdomen soft, very long and slender, extending considerably beyond wing tips; wings elongate with complex net venation; antennae clubbed. Feeble nocturnal fliers; feed on small insects. LARVAE (''Doodlebugs'') (**150**): Small (BL to 5 mm); robust, toadlike insects with conspicuous head, enlarged thoracic region, and outsized sickle-shaped jaws; integument bristly. They lie beneath surface of sandy soil with jaws exposed to catch passing insects. Burrow backwards rapidly. RANGE: Arid and semi-arid parts of the Central Valley, Sierra Nevada, central Coast Range, southern California, and deserts.

SCORPIONFLIES
Order Mecoptera

Scorpionflies derive their name from the bulbous male genitalia, which resembles a scorpion's stinger. However,

California genera do not exhibit this form. These are the most primitive insects with complete metamorphosis. Their most characteristic identifying feature is their long face (head greatly elongated with the antennae situated near the chewing mouthparts). Only 4 relatively rare species occur in the state.

152; Plate 5h. Wingless Scorpionfly. *Apterobittacus apterus*. ADULT: BL up to 20 mm; entirely wingless; legs long and slender; yellow-brown to dull olive green. Feed on (**155**) crane flies and other small insects. LARVA: Caterpillar-like, but with abdominal legs reduced to immovable spines, the back and sides with lobed protuberances. They live in sod

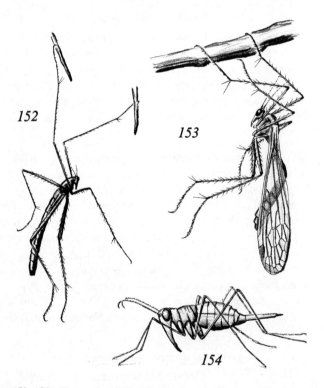

Figs. 152–154 Mecopterans: 152, Wingless Scorpionfly, *Apterobittacus apterus;* 153, Green Stigma, *Bittacus chlorostigma;* 154, California Snow Scorpionfly, *Boreus californicus.*

among grass roots. RANGE: Grassy hillsides, mainly in central Coast Ranges; common in the San Francisco Bay area.

153. Green Stigma. *Bittacus chlorostigma.* This is the only California mecopteran that could be considered common. Large numbers sometimes appear on grassy hillsides in the spring. ADULT: Large (BL 25 mm, WL 25 mm); resembles a large (**155**) crane fly but has 4 equal wings; tarsi hook around perch; a pale green spot near the tip of each wing. Predaceous on various small insects which they capture in flight with their raptorial hind tarsi. RANGE: Southern Sierra Nevada, foothills.

154. California Snow Scorpionfly. *Boreus californicus.* ADULT: Small (BL 4.5 mm); black; male wings only about half as long as the abdomen; female wingless. Appears in winter and early spring (even seen running on surface of snow) in the moist boreal zone where it feeds on mosses. RANGE: Mountain ranges of northern California, south to Nevada County. The **Desert Scorpionfly** *(Boreus notoperates)* is similar in structure but inhabits a considerably more arid life zone, chaparral and yellow pine regions of Mount San Jacinto.

FLIES, GNATS, AND MIDGES
Order Diptera

The Diptera is an important order, but it is not thoroughly studied in California. The presently recognized 5600 species probably represent a considerably smaller number than the actual total fauna. These are minute to medium-sized insects with only one pair of wings. The hind pair is modified into knobbed structures, called halteres, which have a sensory function. Many species are totally wingless (halteres persist), but when present the wings are always membranous and usually transparent with relatively few veins (except in the most primitive groups). The mouthparts are adapted for taking liquid food by lapping, sponging, or piercing and sucking. Metamorphosis is complete and the larva is a legless maggot, grub, or wormlike creature.

There are two major types of Diptera. A primitive group, called midges or gnats (suborder Nematocera), is characterized in the adult stage by long antennae with many similar segments and long, slender legs; the larva has a well-developed head and is frequently aquatic. The adults of the second group, the so-called higher flies (suborder Brachycera), typically possess short antennae in which one or more segments are strongly modified differently from the others, and the legs in the majority of species are short and stout; the larvae are extremely varied in form and habitat (wormlike and grublike types common), but most are typical maggots without a recognizable head. Flies in the suborder Brachycera are referred to as "muscoid" if they bear some resemblance to the common (**206**) House Fly; the especially small, bothersome forms are called "gnats" even though they are not members of the Nematocera.

Because so many Diptera are bloodsuckers or live and develop in filthy materials, this is the most significant order in the transmission of diseases of mammals, including man.

References

Bohart, R. M. and R. K. Washino. 1978. *Mosquitoes of California.* Berkeley: Div. Agric. Sci.; 153 pp.

Carpenter, S., and W. LaCasse. 1955. *Mosquitoes of North America (North of Mexico).* Berkeley and Los Angeles: University of California Press.

Cole, F. R. 1969. *The Flies of Western North America.* Berkeley and Los Angeles: University of California Press.

Eldridge, B. F., and M. T. James. 1957. The typical muscid flies of California. *Calif. Insect Survey, Bull.* 6:1–17.

Huckett, H. C. 1971. The Anthomyiidae of California, exclusive of the subfamily Scatophaginae. *Calif. Insect Survey, Bull.* 12; 121 pp.

———.1975. The Muscidae of California exclusive of the Muscinae and Stomoxyinae. *Calif. Insect Survey, Bull.* 18; 148 pp.

James, M. T. 1955. The blowflies of California (Calliphoridae). *Calif. Insect Survey, Bull.* 4:1–34.

Middlekauff, W. W., and R. S. Lane, 1980. Adult and immature Tabanidae of California. *Calif. Insect Survey, Bull.* 22.

Oldroyd, H. 1964. *The Natural History of Flies.* New York: Norton Co.

Wirth, W. W., and A. Stone. 1956. Aquatic Diptera. In *Aquatic Insects of California,* ed. Usinger, pp. 372–482. Berkeley and Los Angeles: University of California Press.

Suborder Nematocera

CRANE FLIES
Family Tipulidae

Crane flies are small to large, very slender, long-legged flies whose appearance suggests giant mosquitoes (hence the common name, "mosquito hawk"). Their soft mouthparts are not adapted for biting or predation. Because their legs detach easily, the adults are not popular with collectors. The larvae are rather tough-skinned, cylindrical, and gray, brown, or black maggots, often referred to as "leather jackets." Fleshy, finger-like protuberances from the posterior and the protrusile head also identify them. Tipulids are abundant in all kinds of damp habitats; the larvae live in grass roots or decaying vegetation, under bark of rotting logs, or are aquatic. Most feed on decaying organic matter, but some are predaceous or feed on living plants such as mosses. This is the largest family of Diptera, with about 1500 described species in North America; more than 400 are known in California.

Plate 6a. Giant Crane Fly. *Holorusia rubiginosa.* ADULT: Enormous (BL 35 mm); generally reddish brown, with broad, white bands on side of thorax; wings brown. Most commonly seen in spring and summer near streams; frequently attracted to lights at night. LARVA: Very large (BL up to 60 mm), cylindrical, with dark, leathery skin and posterior, fingerlike processes. Semi-aquatic or aquatic; in shore vegetation or trash and on bottom of small stream pools. RANGE: Coast Ranges, Sierra Nevada (lower elevations), and southern California mountains. This is one of the world's largest flies.

155; Plate 6b. Common crane flies. *Tipula.* A genus with many species of familiar insects, 170 known from California. ADULTS (**155**): Large (BL 15–20 mm); generally brown, with clear or brown-tinted wings. Active most of year; commonest in meadows, grassy hillsides, and moist, wooded areas; feed on nectar and other natural solutions. LARVAE (**Plate 6b**): Dark brown or black, thick-skinned, wormlike maggots; fleshy, fingerlike protuberances from posterior. In soil rich with humus, rotting logs, etc.; feed on roots and decaying vegetation, occasionally injurious to plants, bulbs, and tubers. RANGE: General throughout state. One species, the **Range Crane Fly** *(Tipula simplex)* is destructive to grasses, especially in the Central Valley. In seasons when conditions are favorable for the larvae, they denude vast areas of grassland and grain crops.

FUNGUS GNATS
Family Mycetophilidae

156. Fungus gnats are small to medium-sized (BL 5–10 mm), with the body compressed side to side. They are generally red-brown, brown, or yellowish, with long legs, extra long coxae, and large spurs at tips of tibiae. They are most common in dark, damp places such as undercut stream banks. The larvae are slender, wormlike, whitish or yellowish, often with a dark head, transparent skin, and appendages entirely lacking. They live in fungi, decaying wood or vegetation, and moist soil and are frequently gregarious. Mycetophilids are diverse and probably over 200 species are generally distributed in wet areas throughout the state. The majority belong to the genus *Mycetophila (* **156**). A common species is the **Mushroom Gnat** *(M. fungorum),* which is found widely in commercial mushroom cultures.

ROOT GNATS
Family Sciaridae

157. Sciarids are small (BL 2–3 mm), fragile, black gnats, similar to fungus gnats but with shorter legs (coxae much shorter) and kidney-shaped eyes (with a bridge above the antennae). The three simple eyes (ocelli) are conspicuous. Sciarids do not bite but are annoying when they appear indoors (the usual gnat in fresh paint!). The larvae are slender and wormlike; they frequent moist, shady places where they scavenge on decomposing plant matter. Commonly they infest soil and humus mixes used for potted plants. The dozen or so species in California are generally distributed, and no doubt many others await discovery. Most are members of the genus *Sciara (* **157**).

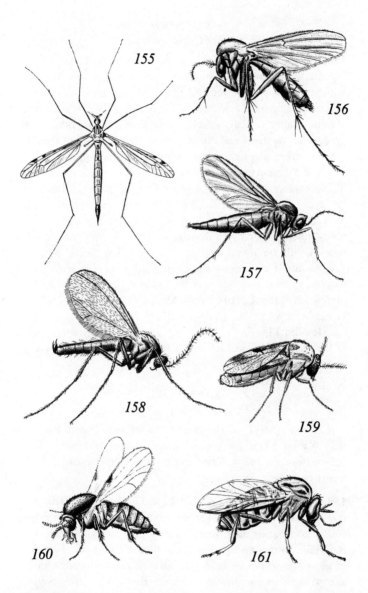

Figs. 155–161 Nematoceran Flies: 155, Common Crane Fly, *Tipula* species; 156, Fungus Gnat, *Mycetophila* species; 157, Root Gnat, *Sciara* species; 158, Hessian Fly, *Phytophaga destructor;* 159; Punkie, *Culicoides* species; 160, Valley Black Gnat, *Leptoconops carteri;* 161, Black Fly, *Simulium* species.

GALL MIDGES
Family Cecidomyiidae

158. A group of small (BL 2–3 mm) gnats, gall midges are especially delicate with very long legs, broad wings (usually with only 3–4 veins reaching the wing margin), and fine, threadlike antennae. Larval habits vary; some feed on decaying organic matter (usually of plant origin) or fungus; others are predaceous on mites and minute insects; the majority of species feed on internal plant tissue and cause deformations (galls) which are characteristic in form and color for each species. These species are most readily recognized by host plant and gall type. This is one of the 2 or 3 largest families of Diptera, with about 100 species known in California, in addition to many others which await discovery. Probably upwards of 400 species will be found to occur in the state. Some notable species are the **Sagebrush Gall Midge** *(Asplondylia artemisiae)*, which causes woolly bladder galls on sagebrush *(Artemisia);* the **Cactus Fruit Midge** *(A. opuntiae)*, whose larvae burrow into the fruits of Prickly Pear Cactus; and (**158**) the **Hessian Fly** *(Phytophaga destructor)*, a major wheat pest in the eastern United States, but fortunately of only sporadic occurrence in this state.

SAND FLIES
Family Ceratopogonidae

Ceratopogonids are tiny gnats (''no-see-ums'') with BL less than 3 mm. They are noticed mostly for their fierce biting habits (females only). The larvae are eel-shaped and live in moist or aquatic habitats.

159. Punkies. *Culicoides.* ADULTS: Tiny gnats (BL 1.0–2.5 mm), usually with mottled thorax and spotted wings (an especially noticeable dark mark midway on anterior margin). RANGE: Numerous species occur throughout state. Females of most bite, some fiercely, and are well known to inhabitants and visitors of the lower Colorado River *(C. variipennis)*, northern coasts near salt marshes *(C. tristriatulus)*, Central Valley *(C. reevesi)*, and northern mountains *(C. obsoletus)*.

160. Valley Black Gnat. *Leptoconops carteri.* ADULT: Tiny gnat (BL 2 mm); shining black with white wings. Very annoying, primarily in April-June. LARVA: Lives in pore spaces in wet clay or adobe soils rich in organic matter. RANGE: Most abundant in western Sacramento Valley. This is a well-known species (referred to formerly as *L. torrens*) with a bad bite that produces a long-lasting, inflammatory swelling. A number of similar species in this genus *(Kerteszi* complex) occur in the state, including biters around coastal estuaries and desert saline lake margins.

BLACK FLIES
Family Simuliidae

161; Plate 6c. Black flies. *Simulium.* ADULTS (**161**): Biting gnats, like sand flies (**159, 160**), but larger (BL up to 4 mm) and distinctively "hump-backed" in appearance; body often shiny gray; legs mottled with yellow; wings broad and transparent. Females bite man and other mammals and constitute a major nuisance when abundant near streams; they are known to attack helpless livestock in such numbers as to cause death. LARVAE (**Plate 6c**): Elongate, swollen rearward, with terminal sucker; head well developed, with large mouth fans. Found in streams, on rocks in swift water, often in massive groups. PUPAE: In cocoons with open, wide end facing downstream. RANGE: More than 20 species; throughout state. Species with bad reputations are *Simulium vittatum* along the Colorado River and generally throughout the state, *S. tescorum* near desert springs and other wet places, *S. trivittatum* along rivers in the Central Valley, and *S. venator* east of the Sierra Nevada.

NET-WINGED MIDGES
Family Blephariceridae

162, 163. Comstock's Net-winged Midge. *Agathon comstocki.* ADULT (**162**): Large midge (WL 12 mm); gray-brown body with clear wings; eyes large and divided into upper and lower sections; wing membrane with creases or wrinkles. Flies in spray below waterfalls or over swift water in streams; also rests under overhanging banks, boulders, and logs near

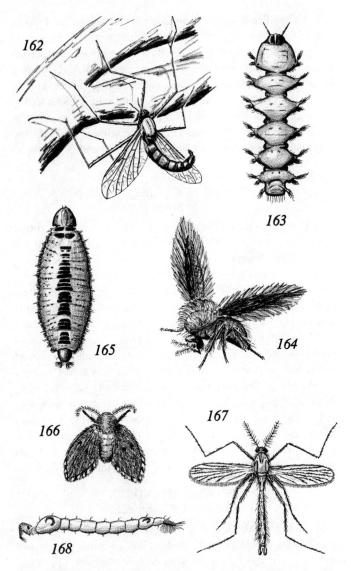

Figs. 162–168 Nematoceran Flies: 162, Comstock's Net-Winged Midge, *Agathon comstocki*; 163, *A. comstocki* (larva); 164, Lance-Winged Moth Fly, *Maruina lanceolata*; 165, *M. lanceolata* (larva); 166, Bathroom Fly, *Telmatoscopus albipunctatus*; 167, Clear Lake Gnat, *Chaoborus astictopus*; 168, *C. astictopus* (larva).

streams; active in spring and summer; female sucks blood of other small flies. LARVA (**163**): Medium-sized (BL 10–12 mm); black; body elongate and divided into 7 sections by strong constrictions; each section with 2 lateral extensions; median series of 6 suckers on underside. Aquatic in shallow, fast-water streams; on smooth boulders and stones. Spring and summer; feed on diatomaceous and algal films on substratum. PUPA: Medium-sized (BL 10 mm); black; oval, hemispherical (with flat underside); a pair of conspicuous, earlike, anterior respiratory horns. Firmly attached (often in groups) to smooth, emergent boulders near waterline or on submerged stones. RANGE: In all granitic ranges throughout the state.

MOTH FLIES
Family Psychodidae

Psychodids are tiny, furry-appearing gnats with wings and body covered with a dense vestiture of scales. Often they perch with the wings held rooflike over the body, or obliquely erect. They resemble minute moths as they scuttle about with a quick, jerky gait. There are about 35 California species.

164, 165. Lance-winged Moth Fly. *Maruina lanceolata.* ADULT (**164**): Very small gnat (WL 1.9–2.5 mm); blackish; wingtips white; wings narrow and sharply pointed. Usually seen crawling (with wings held obliquely erect) on stream boulders or resting under overhanging stones, logs, or vegetation near streams. Late spring and summer. LARVA (**165**): Small (BL 3 mm); black; elongate, flattened; lateral margins lobed; median row of 8 suckers on underside. Aquatic; occur in groups with pupae but more frequently submerged. Feed on diatomaceous and other algal films on substratum. PUPA: Flat, circular, black, disclike, with minute, pointed breathing horns anteriorly; aquatic; firmly attached to smooth, emergent, wet boulders near waterline or on submerged stones. RANGE: Throughout state, in mountain or foothill streams.

166. Bathroom Fly. *Telmatoscopus albipunctatus.* ADULT: Small gnat (WL 2.6–4.1 mm); brown and white scaled; antennae white; white spots at tips of wing veins. Common near damp places; mostly noticed on walls indoors having emerged from drains. LARVA: Very slender, wormlike; BL 8–10 mm; pigmented, with 2–3 dorsal plates on each segment; short posterior breathing tube. Develops in sludgy organic material which accumulates in shallow pools, tree holes, sink traps, drains, etc. RANGE: General throughout state.

PHANTOM MIDGES
Family Chaoboridae

167, 168. Clear Lake Gnat. *Chaoborus astictopus.* ADULT (**167**): Small (BL 3–5 mm); pale yellow; mosquitolike but non-biting, with short mouthparts; scales confined to wing fringe. May emerge in enormous numbers, harrassing lakeshore residents. LARVA (**168**): Transparent, elongate wrigglers similar to mosquito larvae; prehensile antennae directed downwards (used to capture prey); float organs on thorax and abdominal segment 7. Aquatic; member of lake plankton community, migrating upward from bottom during night to feed and take in air from the surface. RANGE: Lakes and reservoirs of north coastal counties; a long-time nuisance at Clear Lake.

MOSQUITOES
Family Culicidae

Mosquitoes are small, slender gnats familiar to almost everyone from the biting and blood-sucking habits of the females. They have long, slender legs, body and wings clothed with scales, and an elongate proboscis fitted with piercing stylets. The antennae are densely featherlike in males, covered with short hairs in females. The aquatic larvae ("wrigglers") and pupae ("tumblers") are both quite active, living in stagnant or slow-moving water, often just beneath the surface film. The wrigglers are elongate, hairy, and have a distinct, movable head, bearing incessantly moving mouth brushes, usually with an elongate, posterior breathing tube; the larvae swim by rapid, lateral jerking movements. The pupae are ovoid, with a suspended, movable abdomen tipped with flat paddles and a pair of breathing trumpets or siphons projecting anteriorly from the thorax. Mosquitoes are the best studied family of Diptera because of their role as carriers of human diseases, such as

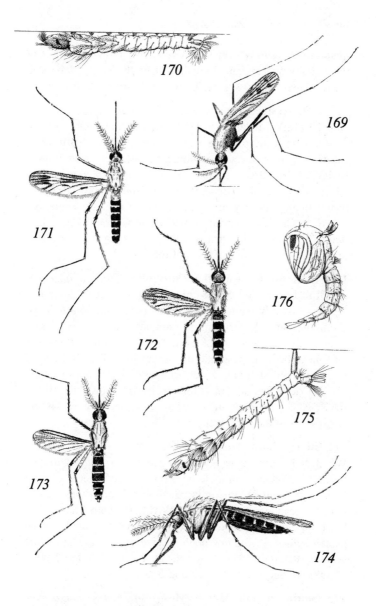

Figs. 169–176 Mosquitoes: 169, Western Malaria Mosquito, *Anopheles freeborni;* 170, *A. freeborni* (larva); 171, Cool-Weather Mosquito, *Culiseta incidens;* 172, Black Snow Mosquito, *Aedes ventrovittis;* 173, Southern Salt Marsh Mosquito, *A. taeniorhynchus;* 174, House Mosquito, *Culex pipiens;* 175, *C. pipiens* (larva); 176, *C. pipiens* (pupa).

malaria and yellow fever. Life cycles are known for most of the world's species and virtually all of the North American forms. There are 47 species known in California.

169, 170. Western Malaria Mosquito. *Anopheles freeborni.* ADULT (**169**): Medium-sized (WL 4.5 mm); abdomen without scales; wing scales forming 4 spots; body position nearly vertical when female bites. Semi-domestic, often entering houses. LARVA (**170**): Breathing tube absent; rests parallel to surface when taking in air. Breeds in clear ground pools (seepage water, rice fields), preferring protection of marshy vegetation. RANGE: Greatest abundance reached in Central Valley; limited in foothills and Sierra Nevada to 2000 m (6500 ft). The principal vector of malaria in California.

171; Plate 6d. Cool-weather Mosquito. *Culiseta incidens.* ADULT: Large (WL 5.5–6 mm); wing speckled; abdominal segments basally white; tarsi with narrow, white basal rings. Common domestic species; most abundant during cooler seasons, bites man readily. LARVA: A pair of conspicuous hair tufts near base of posterior breathing tube. Breeds in wide variety of natural ground waters (ditches, clear pools, swamps) and artificial containers (rainbarrels, tin cans, old tires). RANGE: Common throughout state, from sea level to 3400 m (11,000 ft) in the high Sierra.

172. Black Snow Mosquito. *Aedes ventrovittis.* ADULT: WL 3.0–3.5 mm; black, with long hairs on back of thorax; back of abdomen with metallic, plum-colored sheen; tarsi all dark. Attacks in daytime in large numbers (sometimes clouds) in open meadows; June-July. LARVA: Large hairs on center of head unbranched; hair tuft on breathing tube located beyond pecten spines. In shallow pools after the snow melts; mature rapidly. RANGE: Sierra Nevada at 2100–3350 m (7,000–11,000 ft); Sequoia to Lassen and Shasta counties.

173. Southern Salt Marsh Mosquito. *Aedes taeniorhynchus.* ADULT: WL 2.8–3.2 mm; wing scales all dark; abdominal segments black with white basal bands; proboscis dark with central white ring. Persistent biters during day or night. LARVA: Breathing tube with a single pair of small hair tufts near end; portion of anal segment complete around base of ventral hair brush. Develops in brackish water in salt marshes. RANGE: Along south coast to San Luis Obispo County.

174, 175, 176. House Mosquito. *Culex pipiens.* ADULT (**174**): Medium-sized (WL 3.5–4 mm); reddish brown; tarsi unbanded; proboscis all dark. Highly domestic, commonly entering buildings and biting man. LARVA (**175**): Breathing tube with several pairs of hairs. Breeds in small accumulations of foul water in almost any situation (septic tanks, drains, gutters, ditches, etc.). RANGE: Two subspecies occur in the state: *C. p. pipiens* is northern, mates in swarms, hibernates, and rarely attacks man; *C. p. quinquefasciatus* is southern, usually does not mate in swarms, does not truly hibernate, and bites man readily. Intermediates are found along the coast and in the northern Central Valley. Another *Culex* of medical importance is the **Western Encephalitis Mosquito** *(C. tarsalis)*. ADULT: Medium size (WL 4–4.4 mm); generally dark; tarsi banded; proboscis with median, white band; abdominal segments with black V ventrally. A ubiquitous, semi-domestic species. Bites birds primarily but also man and is the primary vector of Western Equine Encephalitis in California. LARVA: Hair tufts on breathing tube all in a row below the median lines. Found in wide variety of standing, clear (sunny ground pools) or foul water habitats (ditches, excess irrigation water, sewage lagoons). RANGE: General through state below 2700 m (9,000 ft).

WATER MIDGES
Family Chironomidae

There are numerous species of water midges, with about 200 known in California and with many more sure to be found. The adults are generally gnatlike but extremely varied in form and size (BL 0.5–6.0 mm). The wings are long and narrow; legs and antennae long (in males often featherlike); proboscis short, non-biting. Adults often perch with the front legs held in the air and frequently swarm near ponds and lakes, both natural and artificial, causing considerable nuisance. The larvae are long, slender, and wormlike, with a complete head and prolegs on the anterior and posterior body segments. They are mostly white but some possess haemoglobin in their blood and are red ("bloodworms"). The larvae of most species are aquatic, in streams, lake mud, ponds (and even some in marine tide pools), where they dwell in silk cases or nets (lacking in some).

177, 178. Marine Midge. *Telmatogeton macswaini*. ADULT (**178**): BL 3–5 mm, mosquitolike, with long legs; head small and turned down under thorax. Generally gray-brown with silky, brown wings. Crawls on patches of sea lettuce *(Ulva)* on intertidal seashore rocks where it is often covered by surf. LARVA (**177**): Slender (BL up to 15 mm), cylindrical; posterior segment with hooked, stubby, leglike appendages; pale, whitish. Submerged in sandy sediments underlying thick mats

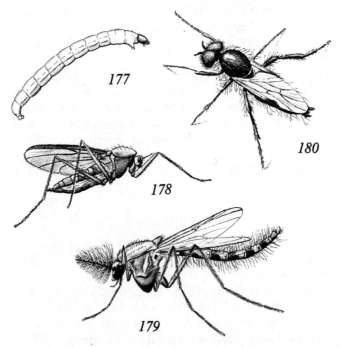

Figs. 177–180 Nematoceran Flies: 177, Marine Midge, *Telmatogeton macswaini* (larva); 178, *T. macswaini* (adult); 179, Common Midge, *Chironomus* species; 180, White-Winged March Fly, *Bibio albipennis*.

of sea lettuce; forms silken tubes incorporating sand grains. RANGE: Rocky portions of entire coast and Channel Islands.

179. Common midges. *Chironomus.* ADULTS: Large gnats (WL 4.5–6.0 mm; BL up to 13 mm); males with broad, featherlike antennae. Sometimes a nuisance near artificial lakes and reservoirs in summer. LARVAE: BL 20–25 mm; body usually curled at both ends, red with haemoglobin; head brown; next to last segment with 4 fingerlike gills. Develop in lakes and ponds, both natural and artificial; often extremely abundant in soft bottom mud where they build a silk-lined tube; respiration is aided by body undulations (irrigating the tube); feed on phytoplankton caught in a silken net spun over the tube mouth. RANGE: General through state.

MARCH FLIES
Family Bibionidae

March flies are small to medium-sized, moderately robust, and often hairy, with shorter legs and wings than most other Nematocera. The eyes are much larger in males, usually meeting at the top of the head, and the 2 sexes sometimes also differ markedly in color and body form. Many species frequent meadows or other areas with abundant decaying vegetation, appearing in spring and fall in great numbers. In some species the males form swarms in which their slow flight makes them easy prey for (**192**) dance flies and other predators. The larvae feed on decaying organic matter at roots of grasses and in leaf mold. They are legless grubs with a well-defined head and often with fleshy processes on the body segments. There are 21 species recorded in California.

180. White-winged March Fly. *Bibio albipennis.* ADULT: Medium-sized (WL 5–9 mm, BL 5–11 mm); shiny black; hairy; with milky white wings; flight weak. LARVA: Small (BL 10 mm); cylindrical maggot; head distinct; skin leathery and warty in appearance; with enlarged posterior pair of spiracles. In soil feeding on plant roots (tubers, grass, etc.). RANGE: Throughout the state except in high boreal zone and deserts, usually near water courses. The **Black Urban Bibionid** *(Dilophus orbatus)* is a smaller (BL 5–7 mm to wing tips), black species, commonly seen flying over lawns during summer and fall; the females are especially conspicuous, having all black wings; this species apparently was introduced from the Gulf States and it was first collected in Los Angeles in 1950 and in the San Francisco Bay area during the 1960s.

Suborder Brachycera

SOLDIER FLIES
Family Stratiomyidae

Soldier flies are mostly medium-sized to large, bristleless, flattened flies somewhat resembling bees or wasps, often with white, yellow, or green markings. They are distinctive in their habit of holding the basal part of the antennae together while the outer parts diverge, like the letter "Y." These flies often occur in numbers at flowers, on vegetation in wet areas, or around outhouses in campgrounds. The larvae are either carnivorous or scavengers, aquatic or terrestrial, and vary greatly in form. The body may be broad and flattened or elongate and tapering to the tail with the last segment drawn out and tubular; the thick, leathery skin is impregnated with calcareous matter. Those larvae living in water or mud hang from the surface film by means of a crown of feathery hairs which closes to trap an air bubble. There are about 65 species of stratiomyids recorded for California.

181. Spotted Soldier Fly. *Stratiomys maculosa.* ADULT: A typical member of the family with black and yellow markings. Moderately large (BL 13–14 mm); eyes hairy; a pair of conspicuous spines on the scutellum. Usually seen in the late spring visiting flowers of shrubs where resemblance to wasps evidently protects it from predators. LARVA: Early stages unknown; larva probably aquatic, living in sluggish or stagnant water rich in vegetation and organic debris. RANGE: Found throughout the state but localized in inland valleys and deserts.

Plate 6e. Black Soldier Fly. *Hermetia illucens.* ADULT: Moderately large (BL 15–20 mm); resembles a wasp in form and behavior: shiny black, with dusky wings, a pair of transparent areas ("windows") at the base of the abdomen, the

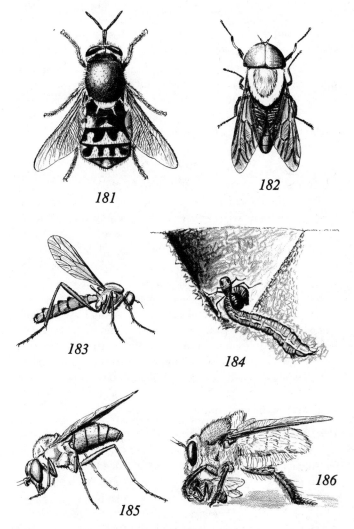

Figs. 181–186 Flies: 181, Spotted Soldier Fly, *Stratiomys maculosa;* 182, Big Black Horse Fly, *Tabanus punctifer;* 183, Worm Lion, *Vermileo* species (adult); 184, *Vermileo* (larva); 185, Biting Snipe Fly, *Symphoromyia* species; 186, Bumble Bee Robber Fly, *Mallophora fautrix*.

median dark line separating these areas simulates wasp waist. Common in summer around habitations (entering houses), stock and poultry ranches, and campground outhouses. LARVA: Elongate (BL 25 mm), tapered toward head, blunt posteriorly; somewhat flattened, with dark brown, leathery skin bearing numerous short bristles. Terrestrial; develops in various organic substances (decaying vegetation, compost piles, garbage, chicken manure) or soil of high organic content where it preys on other fly larvae; may occur in Honey Bee hives. RANGE: Low coastal mountains and valleys, mostly southern California to Bay Area; Central Valley.

HORSE AND DEER FLIES
Family Tabanidae

Tabanids are medium-sized to large, robust flies with short legs. They are powerful fliers with strong wings which are often tinted or spotted. In life the eyes frequently are striped with bright metallic colors. Only the females bite, but like the males, subsist also on nectar and other natural solutions. Their habit of attacking large animals makes them potential vectors of disease, including tularemia and anthrax. The larvae are elongate, cylindrical, tapering at both ends, and usually with bumps or swellings circling the anterior portions of the middle segments. Most are aquatic or semi-aquatic, living in ponds, marshes, or other moist situations such as rotten logs and damp soil, although some live in dry, forest or sand-dune habitats. They prey on other invertebrates. About 75 species of Tabanidae occur in California.

182. Big Black Horse Fly. *Tabanus punctifer*. ADULT: Large (BL 18–22 mm); basically black; male (eyes large, meet on mid-dorsal line) thorax white on rear and side margins; female (eyes smaller, separate on top of head) thorax all pale and hairy; wings opaque; tinted black and with an isolated dark spot on the outer portion. Females feed on blood of large mammals, especially horses; rarely bite man. LARVA: Large (BL 45 mm). In mud and decaying vegetation of marshy ponds or lake and stream margins. RANGE: Throughout state at lower elevations.

Plate 6f. Deer flies. *Chrysops*. ADULTS: Like horse flies but smaller (BL 8–12 mm) and wings usually spotted. Fierce biters and very bothersome. RANGE: Many species throughout state, mountains to desert. Most common in low, swampy areas where many species breed. This genus has numerous species, with about 25 in California. *C. discalis*, with pale, diffuse brown wing markings, is a vector of tularemia. It is found in the Owens Valley and northern Sierra Nevada.

SNIPE FLIES
Family Rhagionidae

Rhagionids are bristleless, slender flies that resemble robber flies. Textbooks say they are predaceous on other insects, but the mouthparts in most genera are small compared to those of robber flies, and predation has been rarely observed. Members of the genus *Symphoromyia* are the most common rhagionids in California and are biters which suck mammalian blood. The larvae are cylindrical grubs with small heads, the body may or may not have false legs. They are carnivorous, preying on other insects in soil, leaf mold, or water. About 40 species of the family occur in California; they include a diversity of forms from the large, hairless, orange *Rhagio* in shaded woods to tiny, golden-haired *Chrysopilus* perching on sunny, streamside vegetation.

183, 184. Wormlions. *Vermileo*. ADULTS (**183**): Small (BL 6 mm); slender; with long legs; brownish gray; abdomen curved under at apex, segments humped dorsally. Rarely seen, rest on vegetation. LARVAE (**184**): Maggot-shaped; BL 8 mm; gray; posterior segment flattened terminally. Frequently mistaken for antlions (**Plate 5g**) because they build funnel-shaped pitfalls in fine sand near rocks to trap ants and other small, crawling insects. RANGE: Sierra Nevada and Transverse ranges at middle elevations.

185. Biting snipe flies. *Symphoromyia*. ADULTS: Small (BL 5–8 mm); wings clear; males generally all brown to gray or gray with orange or light brown abdomen, rather hairy; females hairless, often pale gray with wide palpi and flattened, kidney-shaped, second antennal segment. Common during the spring (lower elevations) and summer in mountainous areas where they bother hikers and campers with their bite; also suck the blood of deer, horses, and other mammals. LARVAE: Early stages little known; larvae develop in damp stream banks and similar places. RANGE: General throughout state, especially mountains and foothills.

FLOWER-LOVING FLIES
Family Apioceridae

Plate 6g. Giant flower-loving flies. *Rhaphiomidas.* ADULTS: Very large (BL 20–30 mm); robust; body yellow, orange, or brown; wings clear, veins in wing apex curve sharply forward to meet margin in front of tip. Noted in spring in desert areas where their large size and extremely fast, buzzing, beelike flight attracts attention. LARVAE: Nothing is known except that larvae are believed to develop in the soil. RANGE: Southern California deserts; occasional in coastal dunes, north to Contra Costa County. *R. trochilus* is restricted to riverine dunes in the Central Valley. It was first found at Merced, collected primarily at Antioch, and is now believed to be extinct owing to destruction of these habitats. The related genus *Apiocera* is more common, with more species, which resemble robber flies with black-and-white patterns but retain the strongly forward-arched veins in the apex of the wing.

ROBBER FLIES
Family Asilidae

Asilids are small to large (BL to 40 mm), elongate, bristly, and sometimes hairy flies with a distinctive head: when viewed from the front, the eyes are strongly raised above the median forehead area. The proboscis is well developed, sharp and rigid, adapted for piercing insect prey. Robber flies are opportunistic predators of other adult insects, taking whatever is available in their habitat. Neither the stings of wasps and bees, distastefulness of milkweed insects, hard shells of beetles, nor insects larger than themselves deter these voracious killers. The victim is seized, often in midflight, and usually is rendered immobile immediately by a stab in the thoracic nerve cord area. Probably over 400 species live in California, the majority of which are small to medium-sized, black, gray, or brown, with clear wings and thin, tapering abdomen. The larvae also are usually predaceous, living in sand, soil, or rotting wood, or in the burrows of other wood-boring insects. They are cylindrical with heads reduced to a small, anterior-pointed structure consisting primarily of mouthparts. In most, the segments of the body have raised or "false legs" circling them.

Plate 6h. Common robber flies. *Stenopogon, Machinus,* and *Efferia.* These genera are representatives of this extensive and diverse family; the majority of conspicuous asilids observed in the state belong to one of them. ADULTS: Medium to large (BL 18–30 mm); bodies bristly; abdomen tapered in females, ending in capsulelike genitalia in males; wings usually transparent; body colors mostly drab brown or gray in *Efferia* and *Machinus* but brown or reddish, often with orange legs, in *Stenopogon* (which may have whitish areas at the base of the wings). All are active predators on other insects, which are usually taken while in flight. They do not attack humans but can inflict a painful bite if handled. LARVAE: Those few known are elongate and cylindrical, with a distinct "head." They live in the soil and feed on other subterranean insects. RANGE: General, most common in arid and sandy areas.

186. Bumble Bee Robber Fly. *Mallophora fautrix.* ADULT: Large (BL 17 mm); robust; densely clothed in yellow and black hairs in a pattern resembling a bumble bee (abdomen yellow, thorax largely black). Rests on vegetation, ready to dart rapidly to catch prey, which consists of a wide variety of other flies, wasps, and bees, including Honey Bees (*Mallophora* are sometimes called "bee killers"). RANGE: Fairly common in southern California at lower elevations. Some members of the robber fly genus *Laphria,* which occur primarily in coniferous forests, also resemble bumble bees but are distinguished from *Mallophora* in being slightly less robust and in lacking a terminal filament on the antennae. Their larvae live in old stumps and logs.

SMALL-HEADED FLIES
Family Acroceridae

Plate 7a. Metallic small-headed flies. *Eulonchus*. ADULTS: Easily recognized by robust shape (BL 5–10 mm); large, humped thorax hiding a small head with long mouthparts; and body coloration of metallic green or blue with bronzy reflections. Usually seen visiting flowers on warm days. LARVAE: Cylindrical maggots tapering at both ends. Internal parasites of spiders. RANGE: Most of the state but most common in forested areas where burrowing spiders are abundant.

BEE FLIES
Family Bombyliidae

Members of the family Bombyliidae are called bee flies because most are densely hairy and beelike in body form. They are small to moderately large with long, slender legs and often a long, projecting proboscis used for probing into deep flowers for nectar. The wings frequently are marked with dark blotches or lacy patterns and are held open, rather than folded over the back, when the insect is perched. The larvae are parasitic on a wide variety of insect larvae, usually those that occupy burrows; hosts include solitary bees nesting in the ground or wood, tiger beetles, and wood-boring beetles. Female bombyliids in some genera are capable of flipping their tiny eggs into open burrows while hovering in mid air. A large number of species is known, over 200 in California.

Plate 7b. Greater Bee Fly. *Bombylius major*. ADULT: Medium-sized (BL 7–12 mm); densely hairy, beelike; usually with black hairs on thorax and 2 lateral spots on the abdomen; anterior half of wings dark brown. Common in all natural areas, particularly sunny spots where wildflowers are abundant; visits deep-throated flowers for nectar, which it takes with long proboscis, and is an important pollinator. LARVA: Grublike, with long body hairs; head poorly developed. Develops in nests of solitary wild bees upon whose larvae it preys. RANGE: Widespread and common, often flying in early spring. A similar-sized species which is also common early in the season is the **Net-veined Bee Fly** *(Conophthorus* *fenestratus).* It has a short proboscis and the wings marked heavily in blackish, irregular network pattern. RANGE: Foothills of the Coast Ranges and Sierra Nevada.

187. Williston's Bee Fly. *Poecilanthrax willistoni.* Large (BL 13–15 mm); body thickly clothed with pale yellow, whitish, orange, and black hairs; sixth abdominal segment almost entirely white; wings with extensive brown pattern; clear areas near apex and mid-way along hind margin. Visits flowers of desert and mountain shrubs; often rests on ground, with wings outstretched. LARVA: BL about 15 mm; body clubshaped, anterior end larger; head partially developed; small breathing pores on first and last segments; skin semi-transparent, light colored, smooth. Internal parasite on cutworms and armyworms. RANGE: Known through most of state, except northwest coast ranges and Central Valley.

THICK-HEADED FLIES
Family Conopidae

188. Thread-waisted Conopid. *Physocephala texana.* ADULT: Medium-sized (BL 9–12 mm); yellow and brown; wasplike, with a large head and narrow waist; anterior portions of the wings shaded in dark brown. Flies low over sandy areas in search of nests of hosts (some accounts say the fly trails the hymenopteran to its burrow); also visits flowers on warm days. LARVA: A medium-sized (BL 15–20 mm) white grub. Parasite on the internal tissues of *Bembix,* sand wasps, and bumble bees. RANGE: Widely distributed in coastal, foothill, and mountainous regions of state; local or absent from humid north coast, deserts, and Central Valley.

HOVER FLIES
Family Syrphidae

Hover flies are so named from their habit of hovering motionless in the air. Flies of the family Syrphidae are small to moderately large, often brightly colored. They bear a resemblance to bees or wasps and almost always are without bristles, although some are hairy and colored like bumble bees. Syrphids also are called "flower flies," because they commonly visit flowers for nectar. Although diverse in size, shape,

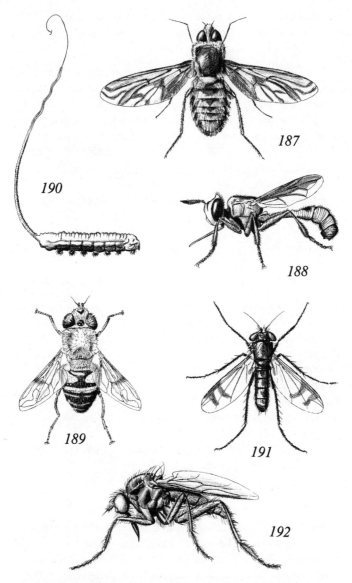

Figs. 187–192 Flies: 187, Williston's Bee Fly, *Poecilanthrax willistoni;* 188, Thread-Waisted Conopid, *Physocephala texana;* 189, Drone Fly, *Eristalis tenax* (adult); 190, *E. tenax* (larva); 191, Garden Long-Legged Fly, *Condylostylus pilicornis;* 192, Common Dance Fly, *Rhamphomyia* species.

and color, all are recognizable by a pigmented line or "false vein" lengthwise along the middle of the wing. Larval habits are varied: some are plant feeders (including bulb maggots); many are predaceous, living externally on plants, especially in aphid colonies where the larvae are cryptically colored in greens and resemble those of lycaenid butterflies; others live in decaying organic matter, such as dung, mud, or stagnant water, rotting wood, or as scavengers in nests of social insects. The larvae are thick maggots with a reduced head. Those living exposed have the hind part of the body blunt, while those living in water have short or long tail extensions ("rat-tailed maggots") used for breathing at the surface. This family of Diptera has numerous species, with probably in excess of 300 in California.

189, 190. Drone Fly. *Eristalis tenax.* ADULT (**189**): A moderately large fly (BL 15 mm); brownish black, light yellow triangles at base of abdomen dorsally; similar to drone Honey Bee in general appearance. Likely to occur almost anywhere, most often noticed visiting flowers. LARVA (**190**) (rat-tailed maggot): A sausage-shaped maggot; with extremely long, extendable, posterior breathing tube. Aquatic; usually in sluggish streams or small, stagnant ponds that are foul with decaying organic matter; also in fresh cattle dung or other fluid animal manure or soil contaminated with same. Extends telescoping breathing tube to surface for oxygen. RANGE: Throughout California. The adult mimics the Honey Bee, which undoubtedly gives it considerable protection. Like its model, it is an introduced species from the Old World. The myth of bees being generated from dead oxen surely derives from development of larvae of *Eristalis* under carcasses.

Plate 7c. Cactus Fly. *Volucella mexicana.* ADULT: Large (BL up to 20 mm); body robust, shiny, purplish black; wing clear, with pigmented areas near base. Spring and summer in arid places; frequent visitors of shrub flowers, especially rabbit brush. LARVA: Large (BL to 30 mm), cylindrical, pale maggot; short, pigmented breathing tube from posterior. Feeds in dead and decaying tissues of large cacti, especially pricklypear and cholla *(Opuntia)*. RANGE: Desert and semi-arid portions of state; Central Valley and Coast Ranges to San Francisco Bay Region.

Plate 7d, Cover photo (upper left). Common hover flies. *Eupeodes, Syrphus, Scaeva, Allograpta, Metasyrphus.* These genera include similar syrphid flies with yellow- and black-banded abdomens. Many species are considered beneficial to agriculture because their larvae are predators of aphids and other plant pests. ADULTS: BL 8–15 mm; thorax shiny and greenish; abdomen largely black with yellow bands; wings clear. Quick in flight, often hovering in mid-air and darting from place to place; common flower visitors and important pollinators. LARVAE: BL 10–15 mm; sluglike, with numerous bumps and protuberances; head end elongate and pointed; usually greenish with median, dorsal yellow line. Move about on leaves and stems, puncturing and grasping aphids or other prey with mouth hooks and sucking out body fluids. RANGE: All portions of state.

LONG-LEGGED FLIES
Family Dolichopodidae

Dolichopodids are small, often metallic green, bristly flies with a curved abdomen and long, slender legs. With only a few exceptions they occur in damp situations, especially along muddy margins of creeks and ponds; several live in the tidal zone of rocky beaches. The flies are predaceous on small, soft-bodied insects. The larvae are also usually carnivorous, living in humus, rotten wood, etc., and a few are aquatic. They are elongate and cylindrical, with a small, retractile "head," and most of the body segments bear false legs equipped with spines used for locomotion. A large number of species live in California, probably in excess of 300.

191. Garden Long-legged Fly. *Condylostylus pilicornis.* ADULT: Small (WL 5 mm); metallic, coppery green, with two incomplete, transverse, dark brown bands across outer half of wing; legs long and slender. Often seen in gardens, walking jerkily and rapidly on leaves. RANGE: Widespread, but primarily a coastal and mountain species; most common at lower elevations.

DANCE FLIES
Family Empididae

Many species of dance flies attract the attention of hikers and trout fishermen because of their habit of gathering in swarms beneath forest trees and over streams. Wheeling and gyrating, these flies perform aerial maneuvers during courtship and mating. Most empidids are predaceous, catching small insects such as weak-flying Diptera, Psocoptera (Corrodentia), and stoneflies. In some species both sexes catch prey, but in most the males do the capturing, form swarms, and present the prey to the female upon mating. The strangest habits are exhibited by a few species in which males secrete material from the digestive tract which is formed into a frothy, white ball or "balloon". This balloon either entraps a small prey or is substituted for prey in the courtship gift-giving sequence. The flies are usually black or gray, small to moderately large, generally slender with a small head, often with one or more pairs of legs and the male genitalia grotesquely modified. The larvae of few species are known. They are cylindrical and spindle-shaped with a small, retractile head, and live in damp places such as humus, decaying wood, or algae at water's edge. Most probably are predaceous. There is a large number of species; about 100 are recorded for California, but many others have not been named.

192. Common dance flies. *Rhamphomyia.* Members of this genus are the dominant empidids in California. ADULTS: Black, gray, or orange; vary greatly in size (BL to 10 mm) and habits. Some visit flowers and apparently do not swarm, while most form small to massive swarms in spring. RANGE: Foothills and mountains to middle elevations throughout the state.

FRUIT FLIES
Family Tephritidae

The common name of fruit flies refers to the fact that larvae of some notorious members of this family feed in fruit. However, most feed in flowers, especially the flower heads of the sunflower family, or they are leaf miners; a few form galls. Tephritids are sometimes called "peacock flies" because males strut about, flicking their outstretched, spotted wings as a part of courtship behavior. The wings of nearly all species are mottled with distinctive patterns of black, brown, or orange. They are most often encountered by carefully observing plants such as woody daisies, rabbit brush *(Chrysothamnus), Baccharis,* and thistles when in bud or early bloom. About 100 species of these flies are known in California. At times a few foreign species, which are serious pests of fruit and other crops, become established in the state, such as Oriental Fruit Fly *(Dacus dorsalis),* Mediterranean Fruit Fly *(Ceratitis capitata),* and Mexican Fruit Fly *(Anastrepha ludens).*

193. Walnut Husk Fly. *Rhagoletis completa.* ADULT: Small (WL 4.5–5.0 mm); thorax brownish yellow; three transverse, dark brown wing bands, the most basal faint, the outer with a broad extension over the anterior wing apex. Active and ovipositing on hosts during summer. LARVA: A small (BL 9 mm), yellowish white maggot. Infests husks of walnuts and affects nut quality by staining husks and kernels. Pupates in soil. RANGE: Indigenous to mid- and south-central United States; introduced into central and western portion of southern California.

KELP FLIES
Family Coelopidae

194. Flat-backed Kelp Fly. *Coelopa vanduzeei.* ADULT: Small (BL 4–6 mm); dark gray; wings clear; body flattened, especially thorax which is wide and depressed. Common (sometimes in swarms) on and under kelp wrack on beaches. LARVA: A small (BL up to 12 mm), cylindrical, white maggot; fingerlike processes on anterior spiracle (segment 2). Develops in moist, decomposing kelp. RANGE: Common along entire coast.

SHORE FLIES
Family Ephydridae

195, 196. Brine Fly. *Ephydra riparia.* ADULT (**195**): Small (BL 4.5 mm), gnatlike; body brown, face yellowish; antennal hair (arista) with long, numerous branches on basal half dorsally; lower portion of face protruding, oral margin wide.

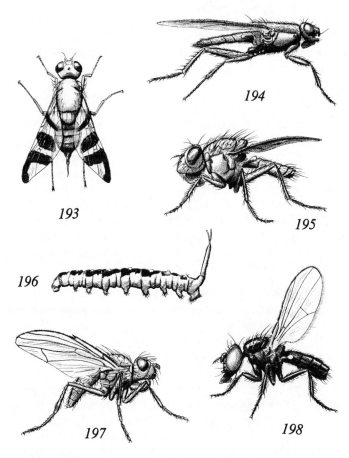

Figs. 193–198 Flies: 193, Walnut Husk Fly, *Rhagoletis completa;* 194, Flat-Backed Kelp Fly, *Coelopa vanduzeei;* 195, Brine Fly, *Ephydra riparia;* 196, *E. riparia* (larva); 197, Koochahbie Fly, *Hydropyrus hians;* 198, Petroleum Fly, *Helaeomyia petrolei.*

Swarms in great numbers on the shores of desert salt lakes and briny marshes (e.g., San Francisco Bay). LARVA (**196**): Medium-sized (BL 10–12 mm); maggotlike; long, apically branched breathing tube posteriorly; many pairs of spined, ventral leglike swellings, the last larger and hooked forward for clasping vegetation. Aquatic, developing in brackish to supersaturated salt water in desert playas or coastal bays. PUPA: Similar in shape to larva but hard and dark brown; attached (often in large numbers) to submerged vegetation or debris. RANGE: Entire coast and southern deserts where suitable aquatic habitats are found.

197. Koochahbie Fly. *Hydropyrus hians.* ADULT: Similar to preceding species; smaller (BL 3–6 mm); black or dark gray, with greenish reflections. LARVA: Maggot-shaped; small (BL 12 mm); whitish, with wide, dorsal pigmented band; ventral leglike appendages with spines, like preceding species; posterior tube short, without basal fork. Lives near bottom of briny and alkaline lakes. PUPA: Like that of preceding species. RANGE: Limited to inland mineral lakes (Clear Lake, Mono Lake, Owens Lake) and ponds. Paiute Indians, in early days, prepared a food (called "Koochahbie") from pupae heaped on lake shores by storms.

198. Petroleum Fly. *Helaeomyia petrolei.* ADULT: Very small (BL 2 mm), gnatlike; shiny black with white balancers (halteres); eyes densely hairy; large bristle of antenna (arista) feathered on upper side; face and oral margin not expanded. Found around natural crude oil seeps and commercial fields. LARVA: Elongate-oval, small (BL 8–10 mm); somewhat transparent; body with 12 segments, pointed anteriorly; 4 projections from posterior end. Live in small pools of oil. Food unknown. RANGE: Known from all the state's petroleum fields and natural oil seeps (e.g., common at Rancho La Brea).

POMACE or VINEGAR FLIES
Family Drosophilidae

Drosophilids are often miscalled "fruit flies" because of the annoying occurrence of a few species around fruit and vegetables in roadside stands, kitchens, and at picnics. However, the larvae feed in decaying fruit and other organic matter, and the common name fruit flies has been universally applied to the flies in the family Tephritidae (p. 161). Pomace flies are small with a somewhat swollen appearance and commonly have red eyes.

199. Vinegar flies. *Drosophila.* ADULTS: Small (BL 2–4 mm); stubby, gnatlike flies; body yellowish to brown, abdominal segments basally banded; eyes bright red. Commonly seen around fruit-processing plants (tomato canneries, wineries, cider mills) or anywhere odor from damaged fruit attracts them. LARVAE: Small (BL to 7 mm); white maggots; posterior spiracles on short stalk. Develop in fermenting or rotting fruit, vegetable, and more rarely, animal matter and fungi, feeding on yeasts therein. Although occasionally a nuisance around homes and in food, *Drosophila* have more than made up for their negative attributes by their usefulness in the study of genetics. No other organism has contributed so much to man's understanding of heredity.

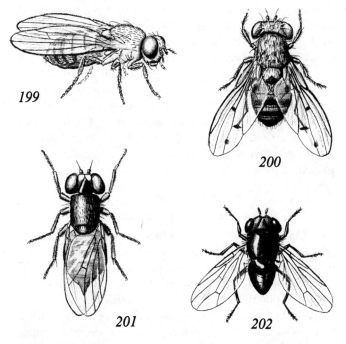

Figs. 199–202 Flies: 199, Common Vinegar Fly, *Drosophila* species; 200, Trail Gnat, *Amiota picta;* 201, Eye "Gnat," *Hippelates* species; 202, Cottony-Cushion Scale Killer, *Cryptochetum iceryae.*

200. Trail "Gnat." *Amiota picta.* ADULT: Small (BL 2–3 mm); generally gray with vague, darker spots on thorax and abdominal segments; dark cloudy spots on wing crossveins, and legs dark-banded. Attracted to humans, especially to the eyes, and may be an extremely irritating pest to hikers, especially on trails near streams. LARVA: Unknown. RANGE: Mountains of southern and central California; Coast Ranges at least to San Francisco Bay Region where it may occur in the valleys as well as upper elevations.

FRIT FLIES
Family Chloropidae

201. Eye "gnats." *Hippelates.* Three important pestiferous species *(H. collusor, H. dorsalis,* and *H. impressus)* in California act as vectors of "pink eye," a bacterial form of conjunctivitis in humans. ADULTS: Small (BL 1.5–2.5 mm); stocky, generally dull or shiny bronze-black. Females attracted to mucous secretions of animals. LARVAE: Minute (BL 2.5–3.5 mm); opaque white maggots. Breed mainly in friable loose soils impregnated with decaying organic matter, including irrigated and tilled agricultural land (citrus groves, orchards—especially date palm—vegetable fields, etc.). RANGE: South of Madera County, below 1850 m (6000 ft) in mountains—including desert ranges; abundant in some localities in San Joaquin, Coachella, and Imperial valleys.

CRYPTOCHETID FLIES
Family Cryptochetidae

202. Cottony-cushion Scale Killer. *Cryptochetum iceryae.* ADULT: Small (BL 1.5 mm); compact; deep metallic blue thorax, abdomen blue-green; wings transparent; eyes extra large; antennae lacking major bristle. Crawl slowly over colonies of Cottony-cushion Scale, into which they insert eggs. LARVA: Stubby, semi-transparent, pear-shaped maggot with 2 long, slender, tail-like processes. Kills the scale by consuming its internal tissues. RANGE: Intentionally introduced in 1888 into California from Australia to control this harmful citrus and ornamental pest. Locally established in several areas of state where host persists.

MUSCID FLIES
Family Muscidae

The family Muscidae includes flies sometimes placed in two other families, the Anthomyiidae and Scatophagidae. Combined, they make up one of the largest families of Diptera, with about 450 species known in California. In many habitats they are our commonest flies. Most are small to medium-sized and dull colored, bearing a general resemblance to the House Fly, but a few are larger and rather brightly colored, as in the Golden-haired Dung Fly *(Scatophaga stercoraria)*. A few species suck blood of mammals, including man, by means of piercing mouthparts (and both sexes bite). The vast majority are incapable of biting and take up nourishment with "sponge"-type mouthparts, visiting all kinds of rotting vegetable matter, foodstuffs, exudates from plants and homopterous insects, plus flowers for nectar. Some species, particularly the House Fly, are implicated in the transmission of diseases such as dysentery and typhoid fever via mechanical pickup from filth; and others cause considerable annoyance by flying around the face, seeking to alight and feed on perspiration and the secretions of the eyes.

There is a great diversity of larval habits: some are plant feeders, as leaf miners or root maggots, and several are important agricultural pests; most species inhabit decaying organic matter and are scavengers, while a few are predators of other soft-bodied larvae in such habitats. All are headless, legless maggots and many are capable of very rapid development in warm weather.

203. Kelp Muscid. *Fucellia costalis.* ADULT: Small (BL 3 mm); dull brown; abdomen with 4 readily visible segments; thorax rounded and bristly. Very common along beaches, often swarming around piles of rotting kelp where they catch and feed upon other small invertebrates, especially "sand fleas" (amphipods). LARVA: A small (BL 3 mm), white maggot. Feeds on decomposing tissue of stranded beach kelp. RANGE: Entire coast, more common in warmer south.

Plate 7e. Golden-haired Dung Fly. *Scatophaga stercoraria.* ADULT: Small to medium-sized (BL 8–9 mm); slender; body

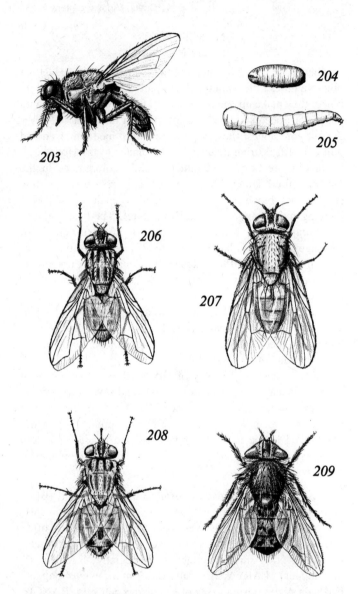

Figs. 203–209 Flies: 203, Kelp Muscid, *Fucellia costalis;* 204, House Fly, *Musca domestica* (puparium); 205, *M. domestica* (larva); 206, *M. domestica* (adult); 207, Little House Fly, *Fannia canicularis;* 208, Stable Fly, *Stomoxys calcitrans;* 209, Cluster Fly, *Pollenia rudis*.

densely covered with bright yellow or tawny hair which is longer in the male. Common in meadows and pastures, often around fresh animal dung; catch other insects (muscids, blow flies) which they feed upon with a hardened, toothed proboscis. LARVA: A medium-sized (BL 15 mm), white maggot. Breeds in all types of fresh excrement (ubiquitous in human feces, cattle and horse droppings) and most decaying vegetation. RANGE: In all parts of state, except the highest elevations; more common where moist.

204, 205, 206. House Fly. *Musca domestica.* ADULT (**206**): Small (BL 5–7 mm); dull gray with 4 darker longitudinal lines on thorax; basal abdominal segments semi-translucent yellow laterally; wings transparent; third vein near tip curving toward second vein; large bristle (arista) of antenna plumose on both sides; soft, spongy proboscis. Man's constant companion, a major pestiferous species in all situations. Feeds on all sorts of substances; directly imbibes fluids and liquifies solids by regurgitation with saliva. LARVA (**205**): White maggot (BL up to 12 mm); posterior breathing pores enclosed in a definite border with sinuous slits; area around spiracles smooth. Breeds in a wide variety of waste substances: animal feed and manures (dog feces important), garbage, and freshly rotting vegetation. RANGE: Literally every place inhabited by man.

207. Little House Fly. *Fannia canicularis.* ADULT: Similar to House Fly but smaller (BL 3–5 mm) and more slender; abdomen yellow basally; the third vein near the tip of the wing curves evenly to margin; abdomen without pattern, gray; large bristle (arista) of antenna bare. On hot, summer days swarms of males often seen hovering, flying in horizontal patterns in the shade of trees, in garages, and other such places. LARVA: BL 6–7 mm; broad and flattened, with prominent lateral and dorsal rows of fingerlike projections. Develops in decaying accumulations of manure, especially from fowl; the species abounds around poultry ranches. RANGE: Of general occurrence in state.

208. Stable Fly. *Stomoxys calcitrans.* ADULT: BL 5–6 mm; very similar to House Fly but with slender, rigid, nonretractable proboscis, having cutting teeth at apex; other differences are the more pronounced and interrupted thoracic stripes and checkered abdomen. Suck blood and seriously torment man and domestic animals. LARVA: Moderate-sized (BL up to 12 mm) maggot; posterior breathing pores sinuous but without encircling border. Breeds in decaying vegetation; often straw and grass near stables, and piled grass clippings and weeds in residential areas. Also develops in seaweed piles on beaches, producing adults which are annoying to bathers ("beach flies"). RANGE: Throughout state, especially in suburban and rural environments.

BLOW FLIES
Family Calliphoridae

Calliphorids are similar to tachinids, muscids, and sarcophagids but are more often metallic green or blue and with more weakly developed bristles, especially on the abdomen. They are most commonly seen at flowers, fresh dung, or swarming around dead animals. The larvae are legless, headless maggots, mostly breeding in carrion, excrement, or other rotting organic matter, usually of animal origin. Blow flies are important from a medical and veterinary standpoint. Some species, including the notorious screwworms *(Callitroga macellaria* and *C. hominivorax,* which are both found in California), feed as parasites or in wound tissue of domestic animals and man or attack nesting birds. Those which frequent contaminated materials often enter houses and can mechanically transmit disease organisms. About 35 species are recorded in the state.

Plate 7f. Green Bottle Fly. *Phaenicia sericata.* ADULT: A beautiful, moderate-sized fly (BL 5–10 mm) with a bright emerald, bluish green, or coppery metallic body; without markings but with numerous black spines; eyes brick red. Extremely common near and inside homes, especially during summer. LARVA: Slender white maggot, tinged with pinkish or purple; BL up to 16 mm; posterior breathing pores pear-shaped with 3 straight, converging slits, in pit surrounded by 10 tubercles. The commonest species in household garbage and urban carrion. RANGE: Known virtually everywhere in state. A related species, *P. pallescens (=cuprina)* may replace *sericata* in coastal and northern parts of the state. Species of **black garbage flies** *(Ophyra)* are also associated with garbage. They are common around refuse dumps in the manure of garbage-fed hogs but are not usually found near homes. They are distinguished from *Phaenicia* by smaller size (BL 5 mm), shiny black color, and white area in the center of the face. The larvae are thick-skinned and have spiny protuberances ventrally. They prey on other insect larvae.

209. Cluster Fly. *Pollenia rudis.* ADULT: Robust; BL 7–9 mm; with checkered black and silvery abdomen; uniquely distinct from relatives by body covering of sparse, crinkly, golden hairs. Moderately common in urban situations; may enter houses in winter in cold climates. LARVA: Medium-sized, smooth maggot (BL up to 12 mm); the last segment with 8 terminal protuberances; terminal spiracles with 3 straight slits, encircled by a weak border. Predaceous on earthworms. RANGE: General throughout state; rare and local in the desert areas.

210. Black Blow Fly. *Phormia regina.* ADULT: Medium-sized (BL 6–11 mm); body shiny blackish or dark olive green; head black; conspicuous orange hair around main thoracic spiracle. LARVA: A large maggot (BL up to 17 mm); creamy translucent white to yellow; posterior breathing pores in pits incompletely encircled by border. Feeds mainly in fresh carrion but sometimes enters healthy tissue of animals from wounds. RANGE: Of general occurrence in the state.

211. Common Blow Fly. *Eucalliphora lilaea.* ADULT: Extremely similar to the European Blue Bottle Fly; identified by slightly smaller size (BL variable but usually 5–9 mm) and heavier upper bristles in row on face lateral to antennae (**211**). Readily enters houses and often attracted in large numbers to human surroundings. LARVA: A typical large maggot (BL up to 18 mm) which cannot be distinguished from that of the European Blue Bottle Fly without examining internal structures. Develops in freshly killed carcasses of small animals,

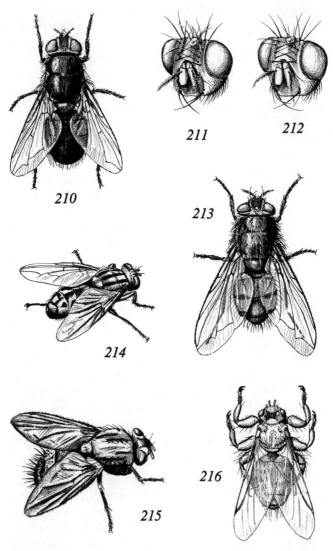

Figs. 210–216 Flies: 210, Black Blow Fly, *Phormia regina;* 211, Common Blow Fly, *Eucalliphora lilaea* (face); 212, European Blue Bottle Fly, *Calliphora vicina* (face); 213, *C. vicina* (adult); 214, Common Flesh Fly, *Sarcophaga* species; 215, Caterpillar Destroyer, *Lespesia archippivora;* 216, Deer Louse Fly, *Lipoptena depressa.*

usually rodents and birds. RANGE: The most abundant and widespread blow fly in California.

212, 213. European Blue Bottle Fly. *Calliphora vicina.* ADULT (**213**): Medium-sized (BL 5–12 mm), heavy bodied; thorax dull blackish, abdomen metallic dark blue with silvery, checkered markings; head as in preceding species but with finer, less extensive upper bristles (**212**); yellow in the mouthpart cavity. Attracted to any foul-smelling, decaying object, especially carrion. LARVA: BL up to 19 mm; spiny bands around abdominal segments; posterior segment tipped with a circle of small tubercles above; two larger lobes below. A very common maggot on dead animals. Occasionally infests healthy animal tissue from infected wounds and from being swallowed. RANGE: Generally distributed in California.

FLESH FLIES
Family Sarcophagidae

Sarcophagids are mostly gray flies with a longitudinally striped thorax and often with a checkered abdomen. They resemble muscids and tachinids, but most species have large bristles only on the hind part of the abdomen, and often the abdominal tip is red. Flesh Flies are most commonly seen at flowers or buzzing about on the ground in dry, rocky areas; in arid places they sometimes constitute an annoyance by persistently alighting on the skin in search of moisture. The larvae of most species occur in dead animal matter or are parasites of other insects, such as grasshoppers and adult beetles; scorpions and snails also serve as hosts. Many of the smaller sarcophagids live on the paralyzed insects provisioned by ground-nesting wasps. Many of the larger species "larviposit" (the eggs are incubated in a pouch inside the female), and the young larvae are placed or dropped from the air onto fresh meat or living insect or mammalian hosts (and sometimes also animal droppings). The larvae are headless and legless, tapered towards the front, with a rounded tail end, and the segments have transverse bands of minute, teethlike spurs. More than 100 species have been recorded in the state.

214. Common flesh flies. *Sarcophaga*. ADULTS: BL up to 15 mm; a little like giant House Flies but seldom enter houses and are distinguished by the 3, instead of 4, black bars on the thorax and by the red-tipped abdomen. LARVAE: Large (BL to 20 mm), pale maggots; lobes present around posterior spiracles, the slits of the latter directed away from the midline. Scavengers on animal feces and carrion or parasitic on other insects. Commonly feed on dead garden snails. RANGE: General.

TACHINID FLIES
Family Tachinidae

The family Tachinidae has more members than almost any other family of Diptera, with probably more than 400 species occurring in California. Many are large and common insects, so that tachinids constitute a conspicuous part of the fly fauna in most natural habitats. They have a well-developed lobe under the mesothoracic scutellum (''postscutellum'') and are often recognizable by numerous heavy bristles covering the abdomen. Tachinids are frequently seen at flowers or buzzing circuitously about in vegetation near the ground. The adults feed on nectar or other plant exudates, while the larvae are parasitic in other insects or rarely other arthropods. Tachinids most often parasitize Lepidoptera caterpillars, but sometimes adult Coleoptera, Hemiptera, and Orthoptera, or larvae of other insects. Females may cement their eggs onto the host integument, often on the heads of caterpillars, but some species deposit large numbers of minute eggs on plants, which are then ingested and hatch within the bodies of the hosts. In other cases, females incubate their eggs and either inject incubated eggs into the host or allow them to hatch and deposit the maggots on surfaces to find their host on their own. Inside the victims, the larvae breathe free air either by perforating the body wall of the host or by a connection to its tracheal system. When full grown, most tachinid larvae leave the host and form puparia on the ground.

215. Caterpillar Destroyer. *Lespesia archippivora*. ADULT: Small (BL 5–7 mm); similar in size to House Fly but with heavy bristles at the tip of the abdomen; third antennal segment very long. LARVA: A thick-skinned, oval maggot. Internal parasite of the caterpillars of many larger moths and butterflies (a perpetual enemy of the Monarch, **312**). RANGE: Widespread at low elevations, including the deserts.

Plate 7g. Spiny Tachina Fly. *Paradejeania rutilioides.* ADULT: Large (BL 16–18 mm); robust; abdomen orange or black (the form occurring in California named *nigrescens*); whole body, especially abdomen, with dense vestiture of enormous, stiff, black bristles; wings smoky. Most abundant in early autumn; often visits flowers, especially of the sunflower family. LARVA: Large, cylindrical maggot. An internal parasite of caterpillars, possibly restricted to larvae of tiger moths (one record from Edwards' Glassywing, **284**). RANGE: Mountains of southern half of state and coastal northern California.

LOUSE FLIES
Family Hippoboscidae

The family Hippoboscidae has only a few members (12 species in California). Louse flies are external parasites of mammals and birds. They are conspicuously flattened and cling tenaciously to their hosts, sucking their blood and sometimes transmitting diseases. The larvae develop one at a time within the female's abdomen, and receive nourishment from internal glandular secretions.

216. Deer Louse Fly. *Lipoptena depressa.* Small (BL 3–4 mm); V-shaped lines on upper surface of abdomen; wings break off after fly has settled on host. HOST: Deer. RANGE: Foothills and mountainous regions of state wherever host found. A related species, *Neolipoptena ferrisi,* lacks abdominal markings and also occurs on deer. Another common species is the **Pigeon Louse Fly** *(Pseudolynchia canariensis).* ADULT: Large (BL to wing tips 8 mm); dull brown; wings retained throughout life, tips folded. HOST: Domestic pigeon. Infests adult birds and squabs. Transmits pigeon malaria. RANGE: Introduced from the Old World; fairly common throughout state on feral birds and in lofts.

RODENT BOT FLIES
Family Cuterebridae

217. Rodent bot flies. *Cuterebra*. ADULTS: Very large, robust flies (BL 20–30 mm), somewhat resembling carpenter bees; body generally black, with contrasting, dense white pubescence on lower sides of thorax, and shiny bluish abdomen. LARVAE: Very large (BL 35 mm); oval grub (rounded at both ends); skin thick with heavy plates and fine spines. Internal parasites of rabbits, wood rats, and mice. RANGE: There are 7 California species, ranging widely with their hosts.

STOMACH BOT FLIES
Family Gasterophilidae

218; Plate 7h. Horse Bot Fly. *Gasterophilus intestinalis*. ADULT (**218**): Medium-sized (BL 11–15 mm); yellowish

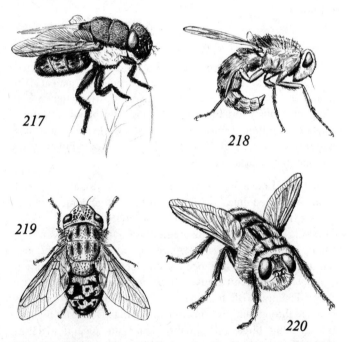

Figs. 217–220 Bot Flies: 217, Rodent Bot Fly, *Cuterebra* species; 218, Horse Bot Fly, *Gasterophilus intestinalis;* 219, Sheep Bot, *Oestrus ovis;* 220, Common Cattle Grub, *Hypoderma lineatum.*

brown; slightly resembles the Honey Bee; thorax completely haired; abdomen rather elongate and frequently curled beneath body. LARVA (**Plate 7h**): Large (BL up to 20 mm); thick-skinned grub; anterior end tapering, posterior blunt; single major row of large, slender, blunt-tipped spines bordering anterior margins of segments. The eggs are laid on hairs of the horse, particularly on backsides of knees, and hatch when licked; the young larvae enter the horse's tongue, eventually working their way down the esophagus to the stomach. RANGE: Throughout area where horses are boarded; common around suburban horse stables.

SHEEP BOT FLIES
Family Oestridae

219. Sheep Bot. *Oestrus ovis*. ADULT: A small bot fly (BL 10–12 mm); generally dull yellow; head yellow with deep pockmarks; thorax with grayish powder and numerous small tubercles, each bearing a yellow hair. Active in hot sunshine; does not feed; female thrusts living maggots up nostrils of host animals. LARVA: BL up to 20 mm; more or less cylindrical, grublike; skin leatherlike, with rows of fine spines ventrally; yellowish, changing to brown when mature. Lives in the nostrils and head sinuses of sheep and goats, causing extreme annoyance and often death. RANGE: Sheep-raising regions of state; also in feral populations of sheep and goats, such as on the Channel Islands.

WARBLE FLIES
Family Hypodermatidae

220. Cattle Grubs. *Hypoderma bovis* and *H. lineatum*. ADULTS: Large (BL 13 mm); robust; with dense, black and light yellow hairs, somewhat similar to a bumble bee; tuft of whitish hairs on sides of thorax, reddish hair at tip of abdomen; longitudinal bald areas on back of thorax. Seldom seen except when ovipositing on animal hosts. LARVAE (Cattle Grubs): Very large (BL up to 30 mm); thick skinned; yellow-brown to dark brown; row of larger spines near anterior margin of some or all segments, patches of small spines near posterior border. Internal parasites of bovine animals; younger stages burrow through deep tissues and lodge beneath skin on the back when mature (''warbles''); hole made for breathing and emergence ruins hides. EGGS: Attached in series *(H. lineatum)* or singly *(H. bovis)* to hairs of the legs or belly, where the larvae hatch and burrow through the hide. RANGE: Found in association with cattle throughout the state.

FLEAS
Order Siphonaptera

Fleas, like lice, have lost all traces of wings as an adaptation to the parasitic life style. They differ from lice structurally in many ways (side-to-side compressed rather than flattened body; jumping hind legs) and in their holometabolous development (passing through larval and pupal stages). Some fleas lack eyes, but the examples described below all have well-developed eyes.

There are 166 species recorded from the state on a variety of birds and mammals, including livestock, pets, and ourselves. Some are potential vectors of plague and typhus.

The larvae are all very similar, white, wormlike forms with long body hairs. They develop in the host's nests, where they feed on organic debris including dried blood passed by the adults.

References

Hubbard, C. 1947. *Fleas of Western North America.* Ames: Iowa State College Press.

221. Ground Squirrel Flea. *Diamanus montanus.* ADULT: Lacks head comb but with thoracic comb; head rounded; labial palpi very long. HOST: Restricted almost entirely to the California Ground Squirrel *(Spermophilus beecheyi)* on which it may be extremely abundant. RANGE: General throughout state except deserts, following range of host. This is an important vector of wild rodent plague and helps maintain the disease in nature. It is dangerous to handle a ground squirrel because, particularly if sick or dying, it may harbor the plague bacilli which can be transmitted by the bite of infected fleas of this or other species, especially the Oriental Rat Flea.

222. Cat Flea. *Ctenocephalides felis.* ADULT: BL 2–2.5 mm; combs of heavy, sharp-pointed dark bristles on lower margin of head and prothorax. HOSTS: Cats, dogs, opossums, rats, humans. Essentially a domestic species. RANGE: Widespread; more abundant along humid, coastal than in drier, inland areas. Particularly abundant in summer. The related **Dog Flea** *(C. canis)* is a much rarer relative of the Cat Flea, known to occur in the San Diego area.

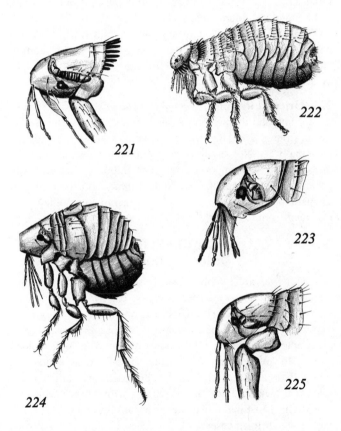

Figs. 221–225 Fleas: 221, Ground Squirrel Flea, *Diamanus montanus* (head); 222, Cat Flea, *Ctenocephalides felis;* 223, Human Flea, *Pulex irritans* (head); 224, Stick-Tight Flea, *Echidnophaga gallinacea;* 225, Oriental Rat Flea, *Xenopsylla cheopis* (head).

223. Human Flea. *Pulex irritans.* ADULT: Lacks head comb and thoracic comb; head outline rounded; large head bristles (below eye). Extremely good jumper. HOSTS: All domestic animals (especially swine) and man; most important flea in farm areas. RANGE: In all parts of the state. *P. simulans,* which is found on deer, is distinguishable only by details in the male but is frequently confused with *irritans.*

224. Stick-tight Flea. *Echidnophaga gallinacea.* ADULT: Small (BL 1–1.5 mm); without head comb or thoracic comb; head angulate, blunt anteriorly; thorax greatly contracted. HOSTS: Wild rodents; domestic on rats and fowl. This flea partly buries its head in the host's flesh when feeding and remains firmly attached for long periods. RANGE: Widespread.

225. Oriental Rat Flea. *Xenopsylla cheopis.* ADULT: Lacks head comb and thoracic comb; head rounded; large head bristles in front of eye. HOSTS: Lives mainly on domestic rodents, principally the Norway Rat, but commonly bites man. This is probably the most important vector of Bubonic Plague in both urban and field situations. RANGE: Limited mostly to seaport towns and proximate inland valleys (Sacramento Valley). Introduced from Old World on rats.

CADDISFLIES
Order Trichoptera

Caddisflies are mothlike insects with long, threadlike antennae. Their wings are membranous, narrow or broad, with many veins, covered with hairs, and held over the body rooflike when in repose. The adults possess mouthparts which may be adapted for licking liquid nourishment with long, folded palpi. However, many species are short-lived as adults and probably do not feed. Most species are nocturnal, dull colored, gray or brown, and are inconspicuous, with a quick, shadowlike flight. Most naturalists are more familiar with the aquatic larvae which form cases.

The eggs are deposited in ponds or streams or on nearby vegetation, and the larvae ("caddisworms") live in water. They have well-developed heads with short antennae and bit-

1a Pacific Spotted Mayfly, *Callibaetis pacificus* (p. 44)

1b Vivid Dancer, *Argia vivida* (p. 47)

1c Big Red Skimmer, *Libellula saturata* (p. 50)

1d Dusty Skimmer, *Sympetrum illotum* (p. 50)

1e Green Darner, *Anax junius* (p. 51)

1f Yellow-banded Stonefly, *Calineuria californica* (p. 52)

1g Hairy Desert Cockroach, *Eremoblatta subdiaphana* (p. 57)

1h Western Subterranean Termite. *Reticulitermes hesperus* (workers) (p. 60)

2a Pallid-winged Grasshopper,
Trimeritropis pallidipennis (p. 65)

2b Orange-winged Grasshopper,
Arphia conspersa (p. 65)

2c Fork-tailed Bush Katydid,
Scudderia mexicana (p. 69)

2d Splendid Shield-back Katydid,
Neduba ovata (p. 69)

2e Jerusalem Cricket,
Stenopelmatus fuscus (p. 70)

2f Tree cricket,
Oecanthus species (p. 72)

2g Mediterranean Mantid,
Iris oratoria (p. 74)

2h Timema, *Timema* species
(pair in tandem) (p. 76)

3a Harlequin Cabbage Bug,
Murgantia histrionica
(p. 89)

3b Western Leaf-footed Bug,
Leptoglossus clypealis (p. 91)

3d Large Milkweed Bug,
Oncopeltus fasciatus (p. 93)

◄ **3c** Western Box Elder Bug,
Leptocoris rubrolineatus (p. 91)

3e Common Milkweed Bug,
Lygaeus kalmii (p. 93)

3f Western Bloodsucking Conenose,
Triatoma protracta (p. 98)

3g Robust assassin bug,
Apiomerus species (p. 99)

3h Woodland cicada,
Platypedia species (p. 113)

4a Red-winged Grass Cicada,
Tibicinoides cupreosparsus (p. 114)

4b Spittlebug, *Aphrophora* species
(nymph) (p. 115)

4c Blue Sharpshooter, *Hordnia circellata* (p. 116)

4d Keeled Treehopper,
Antianthe expansa (p. 117)

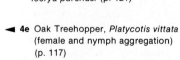

4f Cottony-cushion Scale,
Icerya purchasi (p. 121)

◄ **4e** Oak Treehopper, *Platycotis vittata*
(female and nymph aggregation)
(p. 117)

4g Cochineal scale, *Dactylopius* species
(colony on *Opuntia* cactus) (p. 122)

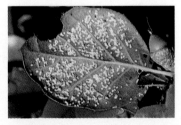

4h Whiteflies, Aleyrodidae
(nymphs on oak leaf) (p. 124)

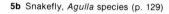

5a California Dobsonfly,
Neohermes californica (p. 128)

5b Snakefly, *Agulla* species (p. 129)

5c Green lacewing, *Chrysopa* species (p. 129)

5d Pacific Brown Lacewing,
Hemerobius pacificus (p. 131)

5e Moth Lacewing, *Oliarces clara* (p. 132)

5f Spotted Large Lacewing,
Polystoechotes punctatus (p. 133)

5g Ant Lion Larval Pits,
Myrmeleontidae (p. 133)

5h Wingless Scorpionfly,
Apterobittacus apterus (p. 134)

6a Giant Crane Fly, *Holorusia rubiginosa* (p. 137)

6b Common crane fly, *Tipula* species (larvae) (p. 137)

6c Black fly, Simuliidae (larvae) (p. 141)

6d Cool Weather Mosquito, *Culiseta incidens* (p. 146)

6e Black Soldier Fly, *Hermetia illucens* (p. 150)

6f Deer fly, *Chrysops* species (p. 153)

6g Giant flower-loving fly, *Rhaphiomidas* species (p. 154)

6h Common robber fly, *Stenopogon* species (p. 155)

7a Metallic small-headed fly, *Eulonchus* species (p. 156)

7b Greater Bee Fly, *Bombylius major* (p. 156)

7c Cactus Fly, *Volucella mexicana* (p. 159)

7d Common Hover Fly, *Allograpta obliqua* (p. 160)

7e Golden-haired Dung Fly, *Scatophaga stercoraria* (p. 167)

7f Green Bottle Fly, *Phaenicia sericata* (p. 170)

7g Spiny Tachina Fly, *Paradejeania rutilioides* (p. 175)

7h Horse Bot Fly, *Gasterophilus intestinalis* (larvae) (p. 176)

8a Silver-striped caddisfly, *Hesperophylax* species (p. 184)

8b Caddisfly, Sericostomatidae (larval cases) (p. 181)

8c Flame Longhorn, *Adela flammeusella* (p. 188)

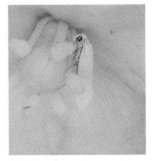

8d Yucca Moth, *Tegeticula yuccasella* (p. 188)

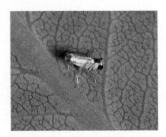

8e Madrone Shield-bearer, *Coptodisca arbutiella* (p. 189)

8f Creosote Bush Bagworm, *Thyridopteryx meadi* (larval case) (p. 191)

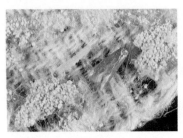

8g Webbing Clothes Moth, *Tineola bisselliella* (adult and larva) (p. 191)

8h Madrone Skin Miner, *Marmara arbutiella* (larval mine) (p. 192)

9a Oak leaf blotchminer,
Cameraria species (p. 192)

9b Locust Clearwing Moth,
Paranthrene robiniae (p. 195)

9c Scythe Moths, Scythrididae (p. 195)

9d Yerba Santa Bird-dropping Moth,
Ethmia arctostaphylella (p. 196)

9e Carpenterworm Moth,
Prionoxystus robiniae (p. 198)

9f Darwin's Green,
Nemoria darwiniata (p. 207)

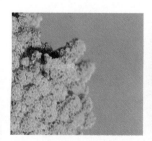

9g Lesser Green, *Synchlora
liquoraria* (larva covered
with flower debris) (p. 207)

9h California Oak Moth, *Phryganidia californica*
(larva) (p. 208)

10a Tussock moth,
Orgyia species
(female) (p. 210)

10b Tussock moth, *Orgyia* species
(larva) (p. 210)

10c California Tent Caterpillar,
Malacosoma californicum
(larvae on tent) (p. 211)

10d Yarn moth,
Tolype species (p. 212)

10e Yellowstriped Armyworm,
Spodoptera ornithogalli
(larva) (p. 216)

10f Bilobed Semilooper,
Autographa biloba (p. 219)

10g Moon Umber, *Zale lunata* (larva) (p. 219)

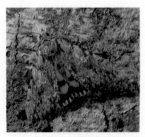

10h Underwing moth,
Catocala species (p. 221)

11a Vestal Tiger Moth,
Spilosoma vestalis (p. 226)

11b Painted Tiger Moth,
Arachnis picta
(adult underside) (p. 226)

11c Painted Tiger Moth,
Arachnis picta (larva) (p. 226)

11d Ceanothus Silk Moth,
Hyalophora euryalus (p. 228)

11e Ceanothus Silk Moth,
Hyalophora euryalus
(larva) (p. 228)

11f Eyed Sphinx, *Smerinthus cerisyi* (p. 231)

11g White-lined Sphinx,
Hyles lineata (adult)
(p. 234)

11h White-lined Sphinx, *Hyles lineata*
(larva) (p. 234)

12a Tomato Hornworm, *Manduca sexta* (larva) (p. 235)

12b Funeral Duskywing, *Erynnis funeralis* (head and thorax of larva) (p. 236)

12c Common Checkered Skipper, *Pyrgus communis* (p. 236)

12d Sandhill Skipper, *Polites sabuleti* (p. 238)

12e Hairstreak, Lycaenidae (larva) (p. 239)

12f Sonora Blue, *Philotes sonorensis* (p. 241)

12g El Segundo Blue, *Pseudophilotes battoides* (p. 241)

12h Purplish Copper, *Lycaena helloides* (p. 243)

13a Lange's Metal Mark, *Apodemia mormo langei* (p. 245)

13b Monarch Butterfly, *Danaus plexippus* (overwintering aggregation) (p. 246)

13c Monarch Butterfly, *Danaus plexippus* (larva) (p. 246)

13d Silverspot, *Speyeria* species (p. 248)

13e Chalcedon Checkerspot, *Euphydryas chalcedona* (p. 248)

13f Satyr Anglewing, *Polygonia satyrus* (p. 249)

13g Buckeye, *Junonia coenia* (larva) (p. 252)

13h Anise Swallowtail, *Papilio zelicaon* (larva) (p. 257)

14a Devil's Coach Horse,
Ocypus olens (p. 271)

14b Rain beetle, *Pleocoma* species
(male and female) (p. 274)

14c Ten-lined June Beetle,
Polyphylla decimlineata (p. 276)

14d Green Fruit Beetle,
Cotinus texana (p. 277)

14e Spotted flower buprestid,
Acmaeodera species (p. 279)

14f Golden Buprestid,
Buprestis aurulenta (p. 279)

14g California Phengodid, *Zarhipis integripennis*
(female consuming millipede) (p. 282)

14h Brown Leatherwing,
Cantharis consors (p. 285)

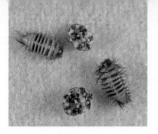

15a Carpet beetle,
Anthrenus species
(adults and larvae) (p. 286)

15b Green Ostomatid,
Temnochila chloridia (p. 290)

15c Green Blister Beetle,
Lytta chloris (p. 293)

15d Convergent Ladybird,
Hippodamia convergens
(hibernation aggregation) (p. 298)

15e Lion Beetle,
Ulochaetes leoninus (p. 302)

15f Red Milkweed Beetle,
Tetraopes basalis (p. 304)

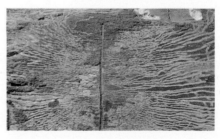

15g Blue Milkweed Beetle,
Chrysochus cobaltinus
(p. 308)

15h Fir Engraver Beetle, *Scolytus ventralis*
(egg and larval galleries) (p. 312)

16a Willow Leafgall Sawfly,
  Euura pacifica (galls)
  (p. 320)

16b Western Orussus,
  Orussus occidentalis (p. 320)

16c Pacific Cuckoo Wasp,
  Chrysis pacifica (p. 329)

16d Red-haired Velvet Ant,
  Dasymutilla coccineohirta
  (p. 332)

16e Mexican Honey Ant,
  Myrmecocystus mexicanus
  (repletes' chamber) (p. 335)

16f Tarantula hawk, *Pepsis* species (p. 336)

16g Sand wasp, *Bembix* species (p. 343)

16h Yellow-faced Bumble Bee,
  Bombus vosnesenskii (p. 353)

ing mandibles. The legs are long, each with a single claw at the tip. The abdomen is soft and caterpillarlike but lacks the false legs of caterpillars. Breathing usually is accomplished by filament gills on the abdomen. Larvae of most species form cases made of silk to which pebbles, plant fragments, or other objects are attached (**Plate 8b**). Usually the cases are tubular, open at both ends with the anterior end larger, and the head and legs protrude when walking or feeding. The form and materials used in case construction are often characteristic of particular species or families. Some species are free-living or form non-movable, gravel-covered retreats. Larvae living in fixed cases are carnivorous and catch their prey in silken snares; those that drag their cases around usually graze on decaying organic matter. The order is a moderately large one, represented in California by about 300 species in 14 families.

The name caddisworm has a curious origin. It is thought to have come from 15th- or 16th-century England, where 'cadise' or 'caddyss' referred to cotton or silk. Itinerant vendors of yarn, braid, and ribbon wore bits of their wares stuck to their coats as advertisement and were called "cadise men." It seems likely that "cadiseworm" was the term applied by anglers to the larvae which form cases covered with bits of debris.

References

Denning, D. G. 1956. Trichoptera. In *Aquatic Insects of California,* ed. Usinger, pp. 237–270. Berkeley and Los Angeles: University of California Press.

Wiggins, G. B. 1977. *Larvae of the North American Caddisfly Genera (Trichoptera).* Toronto and Buffalo: University of Toronto Press; 401 pp.

LONG-HORNED CADDISFLIES
Family Leptoceridae

Leptocerids are common in ponds and lakes, especially in warmer parts of the state, although they are not a dominant form of caddisfly in California. The adults are dark brown to pale green, slender, and possess exceptionally long antennae. The larvae build cases which are usually tubular and slender and covered with fine sand, pebbles, or plant parts. Leptocerids live in permanent, warmer waters, including still ponds and slow-moving rivers. The larvae are distinguished by the relatively long antennae, protruding beyond the front of the head, and by the hind legs which are much longer than the others. Abdominal gills are simple and often few in number. Leptocerids of most genera are indiscriminate feeders, but some feed on freshwater sponges. About a dozen species are known in the state.

226. Log cabin caddisworms. *Oecetis.* ADULTS: BL 8–13 mm to tips of wings; brown with narrow wings and very long antennae, 2× the WL. LARVAE: Probably predaceous; distinguished by having long maxillary palpi and single-blade mandibles; unbranched gills on all but the last 2 abdominal segments. CASE (**226**): Length to 15 mm; covered with short lengths of stems placed transversely; resembling miniature log cabin construction. RANGE: Widespread in mountainous areas.

SNAILSHELL CADDISFLIES
Family Helicopsychidae

227. Northern Snailshell Caddisfly. *Helicopsyche borealis.* This species is the only representative of its family in California; it is widespread and at times abundant in clear streams and at margins of lakes. It has an exceptionally broad temperature tolerance, occurring even in streams with thermal effluents (to 34°C) where no other Trichoptera are found. ADULT: Small (BL 5–8 mm to tips of wings), frail; females larger and darker than males; brownish gray without wing pattern; antennae about as long as FW. LARVA: Spiral-shaped with long legs; sparse gills on abdominal segments 1–5; anal segment with comblike claws. CASE (**227**): Snailshell-shaped, of closely fitted rock fragments; general feeder, grazing on algae, detritus, or animal matter. RANGE: Widespread in coastal areas, foothills, and mountains to moderate elevations. So remarkable is the resemblance of the cases to snailshells, that in 1834 this species was originally described as a snail, having the property of strengthening its shell by gluing on particles of sand!

COLORFUL CADDISFLIES
Family Limnephilidae

The family Limnephilidae contains many of the most conspicuous caddisflies in California. Adults are often large (BL 7–30 mm to tips of wings), and many are orange or reddish brown, sometimes with well-defined wing markings. The larvae live in a variety of habitats. Generally those living in cool

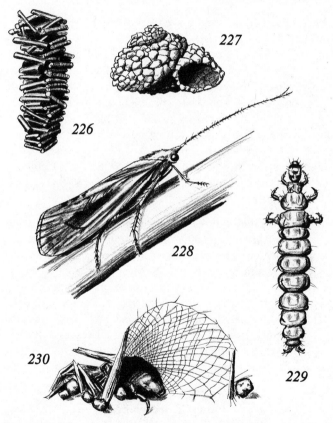

Figs. 226–230 Trichopterans: 226, Log Cabin Caddisworm, *Oecetis* species (case); 227, Northern Snailshell Caddisfly, *Helicopsyche borealis* (case); 228, Reticulated Caddisfly, *Limnephilus* species (adult); 229, Free-Living Caddisfly, *Rhyacophila* species (larva); 230, Webspinning Caddisfly, *Hydropsyche* species (larva within retreat).

mountain streams build cases covered with rock fragments, while those in still, especially warmer, waters use plant materials. Plant tissues are the principal food, although in some genera fungi growing on dead plant matter is the primary attraction, and decaying fish or other animal matter is sometimes eaten. Other limnephilids scrape algae or other organic particles from rock surfaces. More than 40 species have been recorded in California.

Plate 8a. Silver striped caddisflies. *Hesperophylax.* ADULTS: Elongate (BL 17–27 mm to tips of wings); FW orange or rust-colored with a longitudinal silvery streak; abdomen often greenish. LARVAE: Large (BL to 33 mm); rather stout with dark plates on the first 2 thoracic segments and with gills having more than 4 branches on both dorsum and venter of most abdominal segments. CASES: Cylindrical, usually entirely covered with irregularly sized rock fragments, sometimes incorporating small pieces of wood, in streams and at high elevations in still, cold water. RANGE: The 3 California species occur at middle to high elevations in the San Bernardino Mountains, Sierra Nevada, and Cascade Range.

228. Reticulated caddisflies. *Limnephilus.* A related genus whose members make up a more dominant element of the state's Trichoptera fauna, with more than a dozen species in a wide range of habitats. ADULTS: Most are smaller than *Hesperophylax* and are dark rust-colored or orange-brown, often with delicate, lacelike, darker patterns. LARVAE: Primarily live in ponds, lakes, and marshes and build cases of sand grains, pebbles, bark, or wood chips and leaves, arranged lengthwise or transversely. They feed mainly on detritus. RANGE: Widespread, in foothills and mountains.

FREE-LIVING CADDISFLIES
Family Rhyacophilidae

Rhyacophilids are important occupants of cool, fast-moving streams and rivers. The larvae do not construct a portable case or fixed retreat, and they crawl actively over rocks in search of food. Most are predators, although a few eat living or dead plant tissue. At maturity a crude pupal shelter of small stones is constructed on or between rocks. The adults are inconspicuous, small to medium-sized, usually dark brown. About 50 species occur in California.

229. Common free-living caddisflies. *Rhyacophila.* ADULTS: BL 9–17 mm; usually dark brown or reddish brown, broad winged without conspicuous markings. LARVAE (**229**): Variable in form, body with rather strongly hardened integument, usually without filamentous gills on the abdomen, but gills present in some species; anal prolegs each with a strong, curved hook and sometimes an accessory, lateral, curved spike. RANGE: Valleys around the major rivers and in larger mountain streams of the Coast Ranges and Sierra Nevada.

WEB-SPINNING CADDISFLIES
Family Hydropsychidae

Members of the family Hydropsychidae are often abundant in California. The adults swarm to lights in great numbers at mountain localities near streams. Hydropsychid larvae are best known for the elaborate silken nets (**230**) they spin to strain their food from running water. They eat small aquatic insects, algae, or detritus in varying proportions at different seasons. Structurally, hydropsychid larvae are distinguished by having large, dark, dorsal plates of the thoracic segments and branched gills on the underside of all but the first body segment. About 25 species occur in California.

230. Common web-spinning caddisflies. *Hydropsyche.* Often locally extremely abundant, although only 5 species occur in California. ADULTS: Medium-sized (BL 9–15 mm to tips of wings); brown with dull markings of darker brown and whitish; antennae shorter than the wings. LARVAE: Thoracic plates dark, dorsum of abdominal segments also dark patterned; BL to 16 mm. In fixed retreats built of plant fragments and other debris, with a spreading, taut, capture net at the entrance. RANGE: Low-elevation valleys and foothills to middle elevations in mountainous parts of the state.

MOTHS AND BUTTERFLIES
Order Lepidoptera

Lepidoptera are among the most familiar and easily recognized of all insects because of the conspicuousness, large size, and bright colors of many species. Distinguishing characteristics are 2 pairs of wings, usually dissimilar in shape but structurally similar, the wings and body covered with broad, modified hairs (scales) laid down shingle fashion, and a long, coiled proboscis used for sucking fluids.

Moths are less familiar to the general public than butterflies but are far greater in species numbers and diversity of structural modifications and behavioral adaptations. In California about 240 species of butterflies are recorded, representing all 9 North American families of this group. By contrast, no one knows how many kinds of moths occur in the state (many of the smaller ones are not yet described); but there are more than 3,000 species, and these represent 4 suborders and about 50 families.

The metamorphosis is complete; all growth takes place in the larvae (commonly called caterpillars), and the pupae may occur inside silken cocoons (most moths). Adult butterflies and moths may be short- or long-lived, but no change in size or form occurs once they become mature. The adults feed on nectar and other plant secretions, rotting substances, honeydew, or water, although a few species have atrophied mouthparts and do not require nourishment.

The larvae possess strong, chewing mouthparts and feed on a great diversity of substances. Caterpillars of butterflies and most moths eat living vascular plants. Certain moth groups feed in such diverse substances as wood-rot fungi; animal matter including wool, feathers, fur, and bee's wax; fallen leaves; or lichen. Many species are general scavengers, and a few are predators in colonies of other insects. Among the plant feeders, certain species show a pronounced specialization, while others eat a vast array of plants. Lepidoptera are of great economic importance to humans because of the larval feeding; economic species include the Corn Earworm and other cutworms, tent caterpillars, the Mediterranean Flour Moth and other pests of stored foods, lawn moths, the Codling Moth (which causes the 'worms' in apples), Spruce Budworm and other defoliators in coniferous forests, clothes moths, and many others.

Butterflies are dayfliers (although occasionally individuals are attracted to lights at night), and are most easily distinguished from moths by having an enlarged club at the end of each antenna. Moths have threadlike, sawtoothed, or feathery antennae without apical clubs. Most are nocturnal, although some species fly in the daytime and are brightly colored.

Butterflies have no attachment between the wings, which merely overlap, aided by the enlarged basal area of the hindwing. In most moths the wings are attached to each other by a bristle or group of bristles (frenulum) at the base of the hindwing. The frenulum hooks under a flap or tuft of scales (retinaculum) on the underside of the forewing. Usually the sexes are distinguished by the form of the frenulum, which consists of separated bristles in females. The bristles are fused into a single, bristlelike hook in males.

References

Comstock, J. A. 1927. *Butterflies of California*. Los Angeles: publ. by author; 334 pp. + 63 pl.

Emmel, T. C., and J. F. Emmel. 1973. *The Butterflies of Southern California*. Nat. Hist. Museum, Los Angeles County, Sci. Series.

Holland, W. J. 1968 (reprint ed.). *The Moth Book, a Guide to the Moths of North America*. New York: Dover Press.

Howe, W. H. (ed.). 1975. *The Butterflies of North America*. Garden City: Doubleday and Co.

Mitchell, R. T., and H. S. Zim. 1964. *Butterflies and Moths, a Guide to the More Common American Species*. New York: Golden Press.

Moths of America North of Mexico. 1971. London: E. W. Classey. Distributed in U.S. by Entomological Reprint Specialists, Los Angeles. Current series of volumes treating all species of each family in detail, illustrated by color photographs. Published: Sphingidae (R. W. Hodges), Saturniidae (D. C. Ferguson), Bombycoidea (J. G. Franclemont), Pyralidae (parts) (E. G. Munroe), Oecophoridae (R. W. Hodges), and Cosmopterygidae (R. W. Hodges).

Tilden, J.W. 1965. *Butterflies of the San Francisco Bay Region*. Berkeley & Los Angeles: University of California Press. Calif. Natural History Guides 12; 88 pp.

Family Incurvariidae

Members of the family Incurvariidae and the family Heliozelidae (see below) are usually dayflying, small moths, which are distinguished from all other Lepidoptera by having the ovipositor modified for piercing. The females insert the eggs into plant tissue, usually leaves or newly developing seeds, where larval feeding occurs, at least in the early stages.

Plate 8c. Longhorn moths. *Adela.* Species of this genus are among California's commonest dayflying moths in spring. The males usually have longer antennae than the females and are often seen dancing in small swarms above flowery meadows. Both sexes visit blossoms of the sunflower family and other plants for nectar, but females of each species select a particular kind of plant to oviposit in the buds. The **Three-striped Longhorn** *(A. trigrapha)* is our most common species. ADULT: Male black with 3 white stripes across the FW; large eyes nearly covering the top of the head; greatly elongate (15–20 mm), white antennae. Female FW metallic blue with yellowish stripes; HW purplish, head scaling bright orange; smaller eyes; shorter antennae (7–10 mm). BL 5–8 mm to tips of wings. LARVA: At first in flowers of Common and Bicolored Linanthus *(Linanthus bicolor, L. androsaceus);* later in a flat, brown case on the ground, feeding on dead leaves. RANGE: Coast Ranges up to 1200 m (4,000 ft) elevation, Central Valley, and Sierran foothills up to 600 m (2,000 ft). Less common in southern California. A related species, **(Plate 8c)** the **Flame Longhorn** *(A. flammeusella),* has bronze-colored FW, tinged with deep burnished red in some populations; HW metallic purplish. It is abundant in the same areas as *A. trigrapha,* but selects owl's clover *(Orthocarpus)* for depositing the eggs.

Plate 8d. Yucca moths. *Tegeticula.* Females of this genus are distinguished from all other Lepidoptera by having the mouthparts adapted for carrying a ball of yucca pollen. After oviposition into seed-bearing parts of the yucca flower, the moth moves to the flower's stigma and deposits the pollen. The larvae eat only a part of the seed in each pod so that a complete mutual dependence (symbiosis) is shown: the moth is the only pollinator for yucca, and this plant provides the only food used by *Tegeticula* larvae. The **California Yucca Moth** *(T. maculata)* is our most widespread species. ADULT: Rather stout with the wings scarcely exceeding the tip of the abdomen in the female (BL 8–12 mm to tips of wings); white with black specks on the FW in the Transverse Range and northward, entirely black in the San Bernardino Mountains and southward. LARVA: White becoming bright pink or greenish tinged with pink when mature; in green pods of Quixote Plant *(Yucca whipplei)*, leaving in the fall to form cocoons in the ground. RANGE: Monterey County southward and Sierra Nevada from Tulare County southward, wherever the Quixote Plant occurs in natural populations. Two related species are also exclusive pollinators: the **Joshua Tree Yucca Moth** *(T. paradoxa),* a flat, dark gray, sawfly-appearing species on *Yucca brevifolia,* and (**Plate 8d**) the **Yucca Moth** *(T. yuccasella),* a larger, white species on Mohave Yucca *(Y. schidigera).*

231. Bogus yucca moths. *Prodoxus.* Several species of this genus occur in California, each adapted to larval feeding on a particular portion of a yucca's flowering parts, but not in the seeds. The females have no modification for carrying pollen. ADULTS: Tiny (BL 4–8 mm to tips of wings); bronze, gray, or whitish moths, found rapidly scurrying about in yucca flowers. LARVAE: White or pale green, legless grubs tunnelling in pods or main flowering stalks of yucca, where they spend the winter. RANGE: Coast Ranges and deserts, wherever yucca occurs.

Family Heliozelidae

Plate 8e. Madrone Shield-bearer. *Coptodisca arbutiella.* Few people see this tiny moth, but the neatly oval shot-holes left by larvae on leaves of Madrone and manzanita are often noticed. The larvae are seldom abundant in natural situations. Where these plants are used as ornamental shrubs, populations often build up to such high levels that virtually every leaf has many holes. ADULT: Tiny (BL 2.5–3 mm to tips of wings); metallic lead-colored with silvery, black, and orange markings

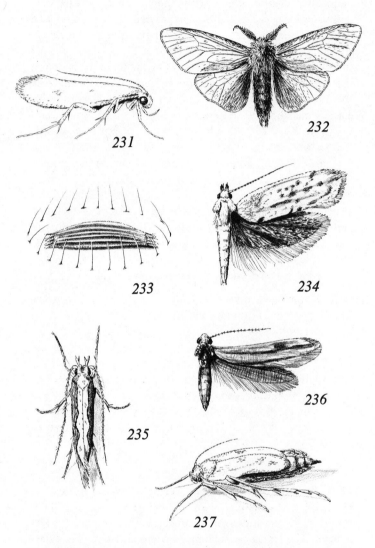

Figs. 231–237 Moths: 231, Bogus Yucca Moth, *Prodoxus* species; 232, Creosote Bush Bagworm, *Thyridopteryx meadi;* 233, Ribbed Casemaker, *Bucculatrix* species (cocoon); 234, Golden Oak Plutellid, *Abebaea cervella;* 235, Diamondback Moth, *Plutella xylostella;* 236, Blackberry Leaf Skeletonizer, *Schreckensteinia festaliella;* 237, Sand-Dune Grasshopper Moth, *Areniscythris brachypteris.*

on the apical half of the FW. LARVA: Flat, forming at maturity in late winter an oval mine between the upper and lower epidermis of the leaf. The floor and ceiling are then cut free and formed into a case in which the larva pupates. RANGE: Coast Ranges from Del Norte to Monterey County and Sierra Nevada foothills. A related species causes the same kind of perforations in leaves of poplar in the Central, Coachella, and Imperial valleys.

Family Psychidae

Members of the family Psychidae are called "bagworms" because the larvae form cases of silk, covered with plant parts, in which they migrate from place to place on the food plants. Females of many species are wingless and never leave the larval case. Psychids are much more abundant and conspicuous in other parts of the United States than in California, where we have only a few species.

232; Plate 8f. Creosote Bush Bagworm. *Thyridopteryx meadi.* ADULT: Male (**232**) winged, stout-bodied, black with semi-translucent wings; BL 10–12 mm. Female wingless, remaining within the case where the eggs are deposited. LARVA: Stout, dark reddish, with a large head and short legs, in a cylindrical, slightly tapering case plastered with Creosote Bush leaves which remain green while the larva is feeding and turn yellow after the moths emerge in late May or June (**Plate 8f**). RANGE: Mojave Desert.

Family Tineidae

Most tineids feed during the larval stage either on living wood-rot fungi, as scavengers in rotting wood, or on fur and feathers in animal nests such as rodent burrows. A few species have adapted to woolens and other man-made animal products, and these are among the most notorious of all Lepidoptera.

Plate 8g. Webbing Clothes Moth. *Tineola bisell iella.* ADULT: Small (BL 5–7 mm to tips of wings); shining golden tan with a roughened tuft of reddish brown scales on the head. Unlike most moths, this and related species are not usually attracted to lights. They retire to dark places so their presence is often unnoticed until larval damage is severe. LARVA: Elongate, white with a brown head; living in silken tunnels webbed together with bits of material upon which they feed. All kinds of fur and wool-containing products are eaten, including clothes, carpet pads, animal skins, and rabbit-foot charms, even the cover of a neglected tennis ball. RANGE: Domestic situations.

Family Gracillariidae

Larvae of the family Gracillariidae are leaf miners, at least during early stages. They feed entirely within one leaf, excavating shallow blotches or linear galleries usually just beneath either the upper or lower leaf epidermis. The form of mine is characteristic for each gracillariid, and most species are restricted in choice to one or a few related plants. There are more than 80 species recorded in California, and undoubtedly many more remain to be discovered.

Plate 8h. Skin miners. *Marmara.* ADULTS: Tiny (BL 2–3 mm); black with thin, white stripes across the FW. LARVAE: Flat, reddish, in elongate, winding mines just under the skin of leaves, stems, or fruit. The mines appear white and persist long after the larvae have abandoned them to form cocoons covered with frothy, silken bubbles. **(Plate 8h) Madrone Skin Miner** *(M. arbutiella)* mines the leaves and stems of Madrone *(Arbutus menziesii)* in early spring. Other species make similar mines on prickly pear cactus *(Opuntia),* bark of roses and berries, or in peels of oranges and apples.

Plate 9a. Live Oak Leaf Blotchminer. *Cameraria agrifoliella.* The large, irregular, whitish blotch mines on the upperside of Coast Live Oak leaves *(Quercus agrifolia)* are often noticeable, especially in urban situations where new growth is produced during late spring and summer. ADULT: Small (BL 2.5–5 mm to tips of wings); golden orange FW, marked with white; perching in a characteristic head downward posture, with the tail end protruding upward **(Plate 9a).** LARVA: Flat, pale greenish, pupating in the mine; the pupa protrudes through the roof of the mine when the moth is ready to emerge. RANGE: Coastal valleys from southern Mendocino

County to San Diego County. This name has been applied to *Cameraria* which are associated with several other species of oak, but these probably will be described as separate species.

Family Lyonetiidae

233. Live Oak Ribbed Casemaker. *Bucculatrix albertiella.* Members of this genus are leaf miners in the early stages of larval growth. Later, the caterpillars feed externally, skeletonizing leaf surfaces. At maturity they form characteristic, white, cigar-shaped cocoons, with longitudinal ribs (**233**), on tree trunks and other objects. When larvae are abundant, the cocoons often are seen first, plastered over walls, garden furniture, and fences. ADULT: Small (BL 3–4 mm to tips of wings); FW checkered whitish and ochreous with a black dot in the middle. LARVA: When mature, dark green with rows of pale spots; grazing on undersides of Coast Live Oak leaves *(Quercus agrifolia)*. COCOON (**233**): White, tapering at both ends with conspicuous, upraised linear ribs (4–5 mm long). RANGE: Coastal valleys from Mendocino County to San Diego. About 30 species of *Bucculatrix* are known in California. The **Hollyhock Leaf Skeletonizer** *(B. quadrigemina)* feeds on plants of the mallow family and is sometimes abundant on garden hollyhocks.

Family Plutellidae

234. Golden Oak Plutellid. *Abebaea cervella.* ADULT: Larger than the following species (BL 9–11 mm to tips of wings); FW shining coppery, golden or brassy colored, sometimes with 2 blackish spots along the inner margin. When the moth is at rest the tips of the wings curve upward, like the outline of a canoe. LARVA: Bright green, strongly tapered toward both ends, with long prolegs; feeding in silken hammocks under leaves of oaks *(Quercus)*. Often the most abundant spring caterpillar on Coast Live Oak. RANGE: Throughout oak woodland areas of the state, up to 1500 m (5,000 ft) in the Sierra Nevada.

235. Diamondback Moth. *Plutella xylostella.* This species probably occurs in more habitats than any other Californian moth. Believed to have been introduced from Europe, the Diamondback Moth is found throughout the United States. It feeds in the larval stage on a wide variety of weedy and culti-vated plants of the mustard family, including cabbage, cauli-flower, radishes, etc. ADULT: Slender, small (BL 6–9 mm to tips of wings); FW brown or blackish, in male with white markings on the inner margin which join in the form of a diamond when the moth is at rest; pattern obscured in female. LARVA: Slender, tapered towards both ends; pale green with yellowish spots; skeletonizing or cutting holes through leaves. RANGE: All of California at a wide range of elevations, from above 2,000 m (7,000 ft) on Mt. Shasta and 3,000 m (10,000 ft) in the White Mountains to the deserts.

Family Heliodinidae

Heliodinids are tiny, dayflying moths with metallic-colored forewings, often marked with bright colors and silvery spots. They perch in a manner unlike any other moth, with the tail end upward and the hind legs held outward.

236. Blackberry Leaf Skeletonizer. *Schreckensteinia festal-iella*. Thought to have been introduced from Europe, this species is widespread in North and Central America, and is often abundant in California. ADULT: Small (BL 4–7 mm to tips of wings); bronze colored with paler and dark linear mark-ings on the forewings. LARVA: Pale green with conspicuous, whitish hairs; feeding on upper or lower sides of Blackberry leaves *(Rubus vitifolius)*, skeletonizing the surfaces, and cut-ting occasional holes. COCOON: A remarkable, oval structure of delicate mesh resembling miniature chickenwire. RANGE: Coastal areas from Del Norte to Ventura counties. The moths fly almost throughout the year.

CLEARWING MOTHS
Family Sesiidae

Sesiids are called clearwing moths or wasp moths because most species have transparent wings, without a covering of scales, and have brightly colored bodies that mimic wasps and bees. They are dayflying and their mimicry has probably evolved through protection from vertebrate predators which learn to avoid stinging insects. Sesiid caterpillars bore into stems and roots, or into galls caused by other insects on plants. About 30 species are known in California, including many forms, colors, and sizes; our largest is the **Manroot Borer** *(Melittia gloriosa)* which lives on manroot *(Marah)* and wild gourd in southern California and reaches 50–60 mm in wingspread, with the HW and immense leg brushes orange.

Plate 9b. Locust Clearwing. *Paranthrene robiniae.* ADULT: BL 13–21 mm; bears a striking resemblance to the Golden Paper-Nest Wasp, with opaque, ochreous or brownish FW and clear HW; thorax black with a yellow hind border, abdomen yellow with broad, black bands on the basal 3 segments. LARVA: Grublike, white with a red-brown head; in dying or recently killed branches of willow, poplar, and locust trees. RANGE: Warmer, low-elevation sites, including the Central Valley, east of the Sierra Nevada, and southern California.

Family Scythrididae

The family Scythrididae includes a large number of small moths which are teardrop-shaped when perched (**Plate 9c**). Many species are dayflying, often black or bronze colored, and visit flowers. Others are nocturnal, usually gray, white, or yellowish species that are attracted to lights. Many scythrids have the peculiar ability to survive death in cyanide killing jars far longer than other insects. The larvae normally are slender, often somewhat colorful, caterpillars that live externally in slight webs on various plants. Probably there are more than 100 species in California, but most are undescribed.

237. Sand-dune Grasshopper Moth. *Areniscythris brachypteris.* This is one of the most remarkable moths in our fauna. Flightless in both sexes, the adults run on open sand dunes like miniature lizards and are able to hop or leap 20–30 times their own length by means of enlarged hindlegs. At night each individual buries itself in a pit dug by alternate swimming-motions of the hind- and forelegs. ADULT: Small (BL 4–7 mm); sand colored, with short wings, FW nearly reaching the end of the abdomen in the male, shorter in female. LARVA: Extremely slender with very short legs, white with paired purplish spots along the dorsum; living in sand-covered, silken tubes attached to stems or leaves of various plants just below the surface of active, moving sand dunes. RANGE: Coastal dunes in the Pismo Beach-Santa Maria area.

Family Ethmiidae

Ethmiids are somewhat slender moths, usually with thin, strongly upcurved palpi. About 25 species are known in California, nearly half of which are dayfliers that are active in winter or early spring when nightly temperatures are too low for small-moth activity. Flight in January, February, or March enables use of annual plants of the borage (Boraginaceae) or water leaf (Hydrophyllaceae) families on which the larvae are specific feeders. Several other species fly in late fall or winter in desert areas.

Plate 9d. Yerba Santa Bird-dropping Moth. *Ethmia arctostaphylella*. ADULT: Medium-sized (BL 10–13 mm to tips of wings); slender, with variably gray forewings having wavy, whitish inner margins; abdomen bright yellow. Characteristically perch along the midrib on uppersides of Yerba Santa leaves and resemble bird-droppings. There are 3 or 4 generations each season; summer moths are paler in color. LARVA: Green or mostly black with orange dorsal spots; in hammocklike webs on leaves of yerba santa *(Eriodictyon)*. RANGE: Foothill areas west of the Sierra Nevada, wherever yerba santa grows, particularly along roadsides. The moth was misnamed because the original collection was a cocoon on manzanita *(Arctostaphylos)*, evidently made by a wandering larva.

Family Oecophoridae

Oecophorids are mostly broadwinged, small to medium-sized, nocturnal moths that have a flat appearance when the wings are folded. The family contains species of diverse appearances and habits. The caterpillars of many feed on living plants, especially of the parsley family, while others are scavengers, living in decaying plant material and sometimes in stored food products. More than 40 species are known in

California, many of which are seldom seen because they are not readily attracted to lights.

238. White-shouldered House Moth. *Endrosis sarcitrella.* This is a common insect in homes and is often mistaken for a clothes moth. ADULT: Small (BL 7–11 mm to tips of wings); FW tan, pointed, variably spotted with blackish; head, thorax, and base of FW white. LARVA: Slender, yellowish white with a yellow-brown head; on dried fruit, meal, and plant detritus, particularly if moldy. RANGE: Urban and rural areas with mild winters. The moths are active throughout the year inside buildings.

Family Gelechiidae

The Gelechiidae is represented in California by at least 200 species in addition to many not yet described. These are mostly small moths characterized by narrow wings, strongly upcurved palpi, and the apex of the HW produced into an abruptly narrowed tip. Caterpillars of gelechiids feed on a great diversity of plants; various species specialize on leaves (including some leaf miners), seeds, twigs, or bark. A few are scavengers in dead leaves or abandoned insect galleries. Both adults and larvae of most species are extremely active, darting or rapidly wriggling away when disturbed. Nearly all gelechiids are nocturnal and they are often attracted to lights.

239. Oak Groundling. *Adrasteia sedulitella.* ADULT: Small (BL 7–8 mm to tips of wings); FW black with upraised scale ridges, often with white markings forming a chevron on the back when the wings are folded. The moths overwinter and may be found in numbers on bark of oaks during summer, winter, or in early spring before the trees have leafed out. LARVA: Bright green including the head, with large black hairs at the tail end; lives in rolls formed of newly developing leaves on various oaks. RANGE: Oak woods at low to moderate elevations throughout California.

GOAT MOTHS
Family Cossidae

Goat moths are moderately large, strong-flying, nocturnal moths. Their caterpillars bore in roots or trunks of woody

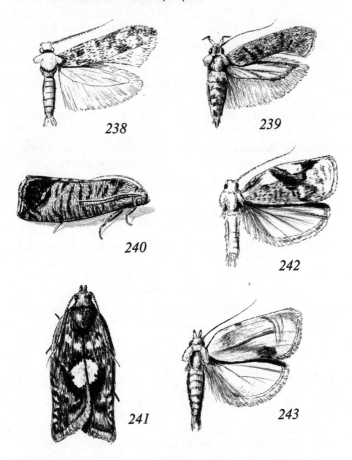

Figs. 238–243 Moths: 238, White-Shouldered House Moth, *Endrosis sarcitrella;* 239, Oak Groundling, *Adrasteia sedulitella;* 240, Codling Moth, *Cydia pomonella;* 241, Variable Oak Leaf-Roller, *Epinotia emarginana;* 242, Fruit-Tree Leaf-Roller, *Archips argyrospilus;* 243, Orange Tortrix, *Argyrotaenia citrana.*

plants. The adults are attracted to lights and they often buzz around on their backs after landing, like clumsy beetles.

Plate 9e. Carpenterworm. *Prionoxystus robiniae.* ADULT: Large, bulky moth (BL to tips of wings 35–45 mm in females; 30–35 mm in males); FW gray, rather sparsely scaled, mottled with black; HW in male orange, margined with black. They fly in April, May, and June. LARVA: Sluggish, grublike, large (BL up to 65 mm); pinkish with dark brown head; in trunks and crowns of Fremont Cottonwood and other hardwood trees including orchard trees; evidenced by large exudations of moist frass and sap, usually at the tree crown or at diseased spots. Larvae probably require 2 or 3 years to reach maturity. RANGE: Low-elevation portions of California, including coastal valleys, Central Valley, and deserts.

Family Tortricidae

Tortricids are small to medium-sized moths, many of which have a rectangular or bell-shaped outline when the wings are folded. There are two subfamilies: the Olethreutinae, with about 250 species in California, and the Tortricinae, represented by about 75 species. Larvae of nearly all olethreutines are borers in seeds, stems, or roots, and most are specialists in host-plant selection. Tortricine caterpillars live as leaf-rollers, and many are indiscriminate feeders on a wide variety of plants. Most tortricids are nocturnal, but a few in each subfamily are dayfliers.

240. Codling Moth. *Cydia pomonella.* One of the most notorious of all Lepidoptera, codling moth larvae are found in apples wherever they are not subject to control programs. ADULT: Small (BL 9–11 mm to tips of wings); somewhat oval in outline when at rest; FW brown mottled with gray with a darker blotch at the terminal end crossed by coppery colored, wavy lines. LARVA: Pinkish, grublike with a brown head; in fruit and seed capsules of apple, quince, walnuts, etc. RANGE: Throughout urban and pome orchard districts of the state. The **Mexican Jumping Bean Moth** *(C. deshaisiana)* is a related species whose larva inhabits the seeds of a native plant of the spurge family *(Croton)* in northern Mexico. After a period of activity, which in nature evidently carries the seed to cracks or other secluded spots, the larva prepares a "trap door" for the moth's emergence and becomes quiescent before pupation. ADULT: Small (BL 9–11 mm to tips of wings), rather robust; FW rectangular dark brown, marked with bluish gray lines. LARVA: Grublike, stout with short legs; body whitish; head brown. Roadside vendors in Mexican border towns often heat the seeds to make the larvae more active at the time of purchase, and this usually kills them; but those packaged by American toy and novelty companies often successfully transform into moths.

241. Variable Oak Leaf-roller. *Epinotia emarginana.* This is one of the commonest small moths in California. The adults emerge in early summer but do not mate and lay eggs until early the following spring. They are often encountered overwintering in aggregations around houses. ADULT: Small (BL 7–9 mm to tips of wings); FW rectangular, having a notch in the end; FW brown mottled with grayish, sometimes with a white patch on the inner margin, or with the inner half ochreous tan or the whole wing tan, with a large black patch on the inner margin. LARVA: Olive green to tan with pale dots and orange-brown head; in leaf shelters on various oaks, manzanita *(Arctostaphylos),* or Madrone *(Arbutus menziesii)* in April, May, and June. RANGE: Throughout forested parts of the state, up to 1500 m (5,000 ft) in the Sierra Nevada.

242. Fruit-tree Leaf-roller. *Archips argyrospilus.* ADULT: Small (BL 9–13 mm to tips of wings); FW dark brown with whitish or silvery markings in male, similar or nearly uniform tan or yellowish tan in female. LARVA: Green with dark olive green back when mature; head brown, mottled with darker markings; in leaf rolls on orchard trees, oaks, and other broadleafed trees. Sometimes extremely abundant on oak, causing defoliation, with larvae dropping to feed on herbaceous plants in the understory. RANGE: Foothills and mountains up to 2000 m (6500 ft) elevation and in the Central Valley.

243. Orange Tortrix. *Argyrotaenia citrana.* Larvae of this species are among the most indiscriminate feeders of all Lepidoptera, eating leaves of all kinds of herbs, shrubs, and trees, even conifers. ADULT: Small (BL 7–11 mm to tips of wings); bell-shaped in outline when at rest; FW dark gray in winter to orange or tan in summer, male marked with variable, diagonal brown bands, female with a dark blotch on the inner margin; HW white. LARVA: Body bright green, head tan; in leaf shelters. RANGE: Coastal plain from Mendocino to San Diego counties and in the Sacramento Valley, especially in urban situations.

PLUME MOTHS
Family Pterophoridae

Pterophorids are called plume moths because the forewings are deeply notched and the hindwings are divided into three linear parts, each with long scale fringes. When perched, the insects roll the forewings around the folded hindwing plumes, resulting in a peculiar sticklike or craneflylike appearance, unlike any other moth. Caterpillars of most plume moths are grublike, covered with short hairs, and colored to blend in with leaves on which they feed. A few species live as borers in bark or buds in the larval stage. More than 30 species are recorded in California; most are nocturnal and attracted to lights.

244. Common plume moths. *Platyptilia.* About 25 species of this genus live in California, each restricted in host choice to one or a few plants. ADULTS: Wing expanse 16–26 mm, generally dark colored; FW in most species with a dark triangular mark on the leading edge just inside the notch. (**244**) **Williams' Plume Moth** *(P. williamsi)* is dark brown with blackish markings. LARVA: Whitish with red bands; in flower heads of Seaside Daisy *(Erigeron glaucus)* and other plants of the sunflower family. RANGE: Beaches and coastal bluffs from Humboldt to Monterey County. A larger pterophorid, the **Morning Glory Plume Moth** *(Oidaematophorus monodactylus)* is a common species around cities; the caterpillars live on Bindweed and other plants of the morning glory family. ADULT: Medium-sized (BL 10 mm; 23–27 mm wing expanse), very slender with long legs; FW ochreous tan to grayish with a dark dot just inside the notch. LARVA: Greenish, covered with short hairs. RANGE: Urban and rural areas of the coastal plain and inland valleys.

Family Pyralidae

A great diversity of species makes up the family Pyralidae. Many have narrow forewings, while in others the forewing is

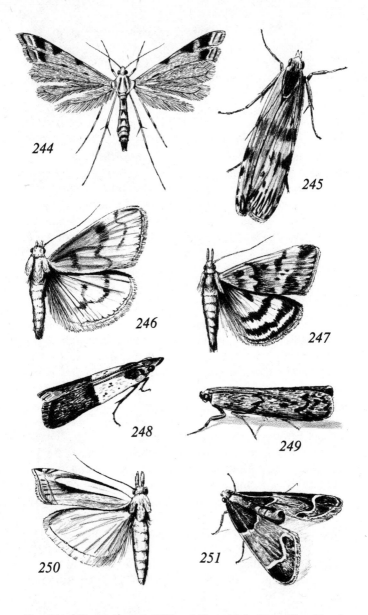

Figs. 244–251 Moths: 244, Williams' Plume Moth, *Platyptilia williamsi;* 245, North American Grass Webworm, *Nomophila nearctica;* 246, False Greenhouse Leaftier, *Udea profundalis;* 247, Weedfield Sable, *Pyrausta sub-sequalis;* 248, Indian Meal Moth, *Plodia interpunctella;* 249, Mediterranean Flour Moth, *Anagasta kuehniella;* 250, Sperry's Lawn Moth, *Crambus sper-ryellus;* 251, Meal Moth, *Pyralis farinalis.*

triangular-shaped. The hindwings are fanlike and held under the forewings when folded. Most pyralids have elongate, snout-shaped palpi, but these structures are shorter and turned up in front of the face in many species. The larval habits are even more diversified, including leaf-feeders, root-borers, moss-feeders, scavengers, predators on other insects and at least one fungus-feeder. One subfamily consists of aquatic species whose larvae feed on submerged plants. Probably 350–400 species have been described from California, including many of our commonest moths, especially in desert regions.

245. North American Grass Webworm. *Nomophila nearctica.* ADULT: Medium-sized (BL 14–18 mm to tips of wings); FW narrow, tan to brown with variable, pale gray streaks and two squarish, brown, partial crossbands. LARVA: Olive-ochreous with a light-edged, dark line along the back and conspicuous black spots; on grasses, clover, and other low-growing plants. RANGE: Throughout California in a wide range of habitats at low elevations, up to 2000 m (6500 ft) in the southern mountains.

246. False Greenhouse Leaftier. *Udea profundalis.* ADULT: Triangular in outline; FW pale rust-orange to brownish with wavy brownish markings; HW whitish with faint brownish markings; BL 10–12 mm to tips of wings. The moths are darker in winter and early spring. LARVA: Slender, pale green with a darker line along the back and two white lines along each side; on underside of leaves of a wide variety of soft-leaved greenhouse plants, weeds, and field crops. RANGE: Widely distributed at low to moderate elevations, coastal counties, Central Valley, and Sierran foothills. A similar species, the **Greenhouse Leaftier** *(U. rubigalis),* is more common in southern California.

247. Weedfield Sable. *Pyrausta subsequalis.* This is one of California's commonest moths in weedy places in urban and rural areas. ADULT: Triangular in outline when at rest; FW brownish, HW orange, banded with blackish; more brightly colored in the female which is smaller, BL 10–15 mm to tips of wings. LARVA: Apparently not described; probably feeds on plantain *(Plantago)*. RANGE: Low to moderate elevations of coastal areas, inland valleys, and Sierra Nevada, from Mt. Shasta to southern California. The moths may be seen at any time of year in coastal areas.

248. Indian Meal Moth. *Plodia interpunctella.* This is probably the commonest stored-food moth in California. Its habits and larval foods are similar to the other pantry moths. ADULT: Small (BL 6–10 mm to tips of wings); FW narrow, grayish white basally with the apical two-thirds brick red. LARVA: Head brown, body pinkish to whitish without discernible body spotting. RANGE: Wherever dried foods are stored; in natural situations in warmer parts of the state.

249. Pantry moths. *Anagasta, Ephestia.* Several species are common inhabitants of stored foods. The larvae eat all kinds of flour, meal, cereals, dried fruits, and nuts, even chocolate and pepper. When households are infested the small, gray moths flutter persistently around kitchens, even after affected foods are removed, because the caterpillars wander to pupate and make cocoons in corners of cupboards and other recesses. The moths may continue to emerge and live for several weeks, so that all new food containers should be stored in closed plastic bags until the moths die out. ADULTS: Small (BL 8–15 mm to tips of wings); FW narrow, rounded, tan or gray, marked with blackish in some species; bodies and HW whitish gray. LARVAE: White or yellowish with a pair of small black crescents on the second and next to last body segments; head brown; in silken webbing on stored foodstuffs. RANGE: Urban and rural situations where dry foods are stored. The largest of these species is the **(249) Mediterranean Flour Moth** *(Anagasta kuehniella),* which has gray FW marked with wavy, blackish cross-bands.

250. Sperry's Lawn Moth. *Crambus sperryellus.* ADULT: Medium-sized (BL 12–16 mm to tips of wings); with long, snoutlike palpi; FW narrow, rectangulate, pale brown with a broad silver streak down the middle. LARVA: Greenish gray or brownish with darker head and conspicuous, paired, black body spots edged with whitish; in silk-lined tunnels in the sod at the base of grasses, on which they feed. RANGE: Coastal plain, Central Valley, and foothills of the Sierra Nevada, in cities where lawns are watered through summer. A related species, the **Western Lawn Moth** *(C. bonifatellus),* has tan or brownish forewings with an obscure, blackish, longitudinal streak and lacks the silver stripe. It is more widespread, occurring in mountain meadows, in the Central Valley, and is more common in cities of southern California.

251. Meal Moth. *Pyralis farinalis.* ADULT: Medium-sized (BL 10–14 mm to tips of wings); with broader, more triangular FW than the other pantry moths; FW brown or reddish at the base and tip, crossed by wavy, whitish lines, pale brown or tan in the middle; HW pale mottled with reddish. LARVA: Dirty whitish, head and first body segment brown; in silken tubes plastered with debris; food includes all kinds of vegetable products, damp material preferred. RANGE: Urban and rural areas.

Family Geometridae

Caterpillars of the family Geometridae lack the first 2 or 3 pairs of false body legs, and as a result they walk in a looping fashion, advancing by measured steps equal to the distance between their thoracic legs and tail end. They are familiar to most naturalists as ''loopers'' or ''inchworms.'' The moths are relatively slender-bodied with broad wings, often with angulate margins. In most species the wings are held out at right angles to the body, appressed to the substrate when the moths are at rest, rather than folded over the back as in most moths. A few geometrids perch with the wings held together, butterfly-style. This is a large family, with more than 300 species in California. There is a wide range of sizes and colors, including a few that are dayflying moths.

252. Oak Winter Highflier. *Hydriomena nubilofasciata.* One of the best insect indicators of the approach of spring, this species appears in great numbers around oaks and at lights in late January and February. ADULT: Medium-sized (24–30 mm wing expanse) with rounded wings; FW gray with variable pattern of olivaceous and dull red-brown, wavy cross-bands.

LARVA: Stout, cutwormlike; yellowish with blue spots to mostly bluish; in leaf rolls on Coast Live Oak *(Quercus agrifolia)* and other oaks. Sometimes the commonest of the spring defoliators of oak. RANGE: Coastal valleys and Coast Ranges from Mendocino County to San Diego.

253. Fragile Gray. *Anacamptodes fragilaria.* ADULT: Moderately large (26–40 mm wingspread); pale gray moth with wavy dark lines across the wings. Seen at lights in southern California throughout the year, most common in fall. LARVA: Cream or pinkish to maroon with darker markings, fifth body segment has 4 triangular tubercules; on various plants.

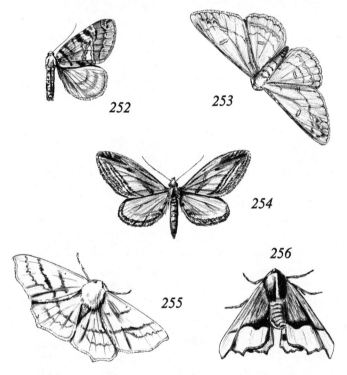

Figs. 252–256 Geometrid Moths: 252, Oak Winter Highflier, *Hydriomena nubilofasciata;* 253, Fragile Gray, *Anacamptodes fragilaria;* 254, Pug Looper, *Eupithecia* species; 255, Omnivorous Looper, *Sabulodes aegrotana;* 256, McDunnough's Leafwing, *Pero macdunnoughi.*

RANGE: Coastal from the San Francisco Bay area to San Diego and inland portions of southern California, including the Colorado River area.

Plates 9f, 9g. Greens. Subfamily Geometrinae. Moths of this subfamily are unusual because most of them are green. The larvae are remarkable in that there are 3 tribes: those that are twiglike in form; those that have large processes on the body, not specialized for attachment of plant matter; and those that have processes on the body bearing special hooks for the attachment of plant fragments. The last are particularly interesting because feeding often occurs in flowers, and petal fragments are attached, producing an extremely cryptic concealment (**Plate 9g**). There are about 25 species of greens in California.

Plate 9f. Darwin's Green. *Nemoria darwiniata.* ADULT: Medium-sized (wing expanse 22–33 mm); wings dull green, each with a reddish dot in the middle with thin, whitish, transverse lines and edged in white, pink at the FW tip; abdomen with 3 white spots surrounded by reddish. LARVA: Yellowish, mottled with brown and lavender; body with surface rugose and with prominent, keel-shaped processes on abdominal segments 2, 3, and 4; a general feeder, reared from willow and Madrone in British Columbia, oak and *Ceanothus* in southern California. RANGE: Coastal areas, Coast Ranges, Sierra Nevada, and southern mountains. Another species, which is similar but lacks the reddish dots in the middle of the wings, is the **Pink-margined Green** *(N. leptalea).* ADULT : Wing expanse 22–30 mm; bright green, wings with 2 transverse white lines, many fine, incomplete whitish lines, and the margins pink. LARVA: Dead-leaf brown with lateral lobes on abdominal segments 2–4, slender and notched; probably a general feeder, reared on buckwheat *(Eriogonum)* and on Christmasberry *(Heteromeles)* in captivity. RANGE: Coastal areas and Coast Ranges from Napa County southward to San Diego. Both species are seen at lights in cities throughout the year.

254. Pug loopers. *Eupithecia.* About 50 species of these small moths occur in California and are often seen at lights.

They are recognized by their narrow wings which are held out flat against the substrate when the moths are at rest. ADULTS: Small to medium-sized (wing expanse 12–35 mm); gray to whitish with darker thin lines across the wings. LARVAE: Variable in form and color, resembling flowers in which they live, such as yellow with brown, dorsal, arrowhead markings in oak catkins or lavender with darker purplish spots in flower-heads of *Ceanothus*. RANGE: Throughout most of the state in a wide range of habitats, except the low deserts.

255. Omnivorous Looper. *Sabulodes aegrotata*. This is the commonest garden inchworm in California and feeds on a wide variety of herbs, shrubs, and trees. ADULT: Wing expanse 40–48 mm, pale tan with 2 irregular, brown bands and faint striations across the wings, or rarely with variable amounts of broad, brown banding to entirely light chocolate brown. LARVA: Yellowish, green, or pinkish with parallel, dark lines along the sides; BL 35–45 mm. PUPA: White with brown antennal cases, in a fluffy cocoon in the leaves. RANGE: Coastal areas, Coast Ranges, and inland valleys, from Humboldt County to San Diego, especially in cities.

256. McDunnough's Leafwing. *Pero macdunnoughi*. Members of the genus *Pero* are heavy-bodied compared to most geometrids and perch with wings curled, resembling dead leaves. They often appear at lights late in the evening when it is too cold for most other moths. ADULT: Wing expanse 36–44 mm; wing color variable, from reddish brown to gray brown. LARVA: Wood brown; in exact simulation of a twig, the cater-pillar assumes a rigid position at right angle to the branch grasped by its tail-end feet; on buckwheat *(Eriogonum)* and probably various other plants, including ornamental garden shrubs.

Family Dioptidae

257; Plate 9h. California Oak Moth. *Phryganidia califor-nica*. This moth often becomes extremely abundant, the larvae completely defoliating live oaks *(Quercus agrifolia* and *Q. dumosa)*. Also commonly found on Cork Oak *(Q. suber)* where it is grown as an ornamental tree in southern California.

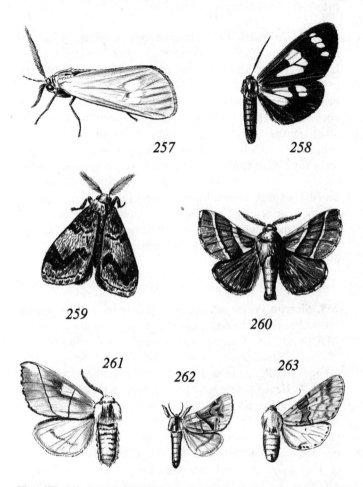

Figs. 257–263 Moths: 257, California Oak Moth, *Phryganidia californica;* 258, Sierran Pericopid, *Gnophaela latipennis;* 259, Tussock Moth, *Orgyia* species (male); 260, California Tent Caterpillar, *Malacosoma californicum* (male); 261, Tawny Prominent, *Nadata gibbosa;* 262, Willow Nestmaker, *Ichthyura apicalis;* 263, American Pussmoth, *Cerura* species.

ADULT (**257**): BL 14–20 mm to tips of wings; pale, dull brown with slightly darker veins; a slow-moving dayflier. LARVA (**Plate 9h**): BL 20–30 mm; olive green with black and yellow lines along the back and sides and a reddish head; on oaks. PUPA: Yellowish with black or bluish markings; lacking cocoon, hanging like a butterfly chrysalis from tree trunks and other objects. All stages are conspicuous and probably are distasteful to birds and thereby protected. RANGE: Coastal areas from Napa County to San Diego. There are 3 generations each year, with successive buildup of numbers in the summer broods, so that extreme defoliation usually occurs only in summer. In central California the populations spread inland and feed on deciduous oaks in summers of outbreaks, but the species is unable to survive freezing in winters and dies out except in coastal areas.

Family Pericopidae

258. Sierran Pericopid. *Gnophaela latipennis.* This moth is often observed fluttering weakly around mountain meadows. ADULT: BL 25–32 mm to tips of wings; black with yellowish white, semitransparent spots on the wings; front of thorax orange on the underside; males have feathery antennae. LARVA: Head orange; body reddish black with stripes of connected, lemon yellow diamonds down the back and on each side above the legs; each segment with raised, blue-black tubercles bearing long hairs, like a sparsely clad woolly bear; on mountain blue bells *(Mertensia),* Grand Hound's Tongue *(Cynoglossum grande),* and California Stickseed *(Hackelia californica).* PUPA: Shiny black with yellow spots on the back; suspended hammocklike from hooks at both ends, in a flimsy cocoon. RANGE: Sierra Nevada, Cascade, and North Coast ranges, at middle elevations.

Family Lymantriidae

259; Plate 10a, 10b. Tussock moths. *Orgyia.* Several members of this genus inhabit California, including one restricted to the immediate sea coast and another in coniferous forests. The moths are nondescript, particularly the females which are wingless, but the caterpillars are brightly colored, have conspicuous hair tufts, and are often seen in gardens. ADULTS: Males (**259**) are medium-sized (26–34 mm wingspread), with rounded brown wings and feathery antennae; FW with gray crossbands and a white spot near the outer, lower corner. Females (**Plate 10a**) lack wings; pale tan, fluffy blobs found clinging to the hairy cocoons. LARVAE: (**Plate 10b**): Gray with yellow and red spots, long, black hair tufts on the front and tail ends, and 4 thick, white tufts on the back, resembling a toothbrush. RANGE: Coastal and mountain areas throughout the state. The common garden species in California was long known as *Hemerocampa vetusta,* but careful taxonomic study probably will show that it should be called *Orgyia gulosa,* with *O. vetusta* restricted to seacoast dune habitats. The **Douglas Fir Tussock Moth** *(O. pseudotsugata)* is a defoliator of conifers in the northern part of the state.

Family Lasiocampidae

260; Plate 10c. Tent caterpillars. *Malacosoma.* Adults of the 4 species in California are dull colored and not conspicuous, but the large, silken tents in which the colonies of caterpillars live are familiar sights on shrubs and trees. ADULTS: Tan to reddish brown, FW usually with contrasting dark or pale lines or bands. Males are medium-sized (wing expanse 22–34 mm) with narrow, feathery antennae. Females are larger (30–38 mm) with threadlike antennae, bulky bodies, and often more uniform-colored wings. LARVAE: Brown or black, usually with conspicuous stripes and chainlike markings of white or bright blue along the back and sides; covered with long, sparse, red-brown or whitish hair. RANGE: Throughout broadleafed forest and chaparral areas of the state, including the Mojave Desert. (**260; Plate 10c**) **California Tent Caterpillar** *(M. californicum)* is the most widespread species in California. The larvae feed in oak, willow, fruit trees, and various shrubs including *Ceanothus* and antelope brush *(Purshia).* ADULT: Males are red-brown with yellowish cross-bands on the FW; HW darker. Females are tan with pale, red-brown bands. Both this species and the **Pacific Tent Caterpillar** *(M. constrictum)* are common on oaks during spring in the Coast Ranges and Sierra Nevada. Adults of *M. constrictum* are colored just the reverse: males tan, females reddish brown. Larvae of *M. constrictum* in these regions usually have conspicuous blue markings, and those of *M. californicum* do not.

Plate 10d. Yarn moths. *Tolype.* ADULTS: Medium-sized (BL 15–19 mm); with broad, rounded wings having whitish or gray with dark cross-bands and fine white lines along the veins; body densely hairy, like a ball of fluff or yarn, white and gray with a black or steel bluish scale ridge on the back of the thorax and base of the abdomen. When at rest the wings are held forward, along the sides of the body, and the abdomen extends beyond them. Males are smaller and have feathery antennae that are broad basally. LARVAE: Gray with faint longitudinal lines; third segment with a black, velvety band and 2 or 3 raised, white spots; body flat underneath; with lateral flaps, each with many long hairs, forming a fringe which aids in shadow elimination and cryptic concealment as the caterpillar rests by day appressed to a similarly colored branch; on conifers, or on mountain mahogany *(Cercocarpus)* and probably other broadleafed shrubs and trees. RANGE: Foothills and mountains to middle elevations throughout the state. Probably there are several species, including one in the Santa Cruz Mountains associated with pine and one in southern California that feeds on hardwoods.

PROMINENTS
Family Notodontidae

Notodontids are medium-sized, stout, nocturnal moths with thick coats of scales and hairs, and usually with reduced mouthparts. Their caterpillars are not hairy but often bear ornate humps or filaments, and they hold the tail end up when at rest. Compared to the eastern United States, the family is poorly represented with about 20 species known in California.

261. Tawny Prominent. *Nadata gibbosa.* ADULT: BL 24–30 mm to tips of wings; bulky, densely hairy moth; tan to brownish, FW with 2 brown-edged, pale lines on either side of a small, double, white spot. LARVA: Head and body green, whitish on the back, with a narrow, yellow stripe along each side; on oak and other broadleafed trees. RANGE: Coast Ranges and Sierra Nevada foothills to about 1200 m (4,000 ft). A related species, the **Yellowstriped Caterpillar** *(Datana perspicua),* is about the same size but has a deep cinnamon red thorax and pale tan FW crossed by 4 thin, red-brown lines; HW whitish. LARVA: BL 50–60 mm; head black, body dark reddish brown or black, with bright yellow stripes along the back and sides, and clothed with sparse, white hair; on broadleafed forest and orchard trees. RANGE: North Coast Range, Sierra Nevada, and mountains of southern California up to 900 m (3,000 ft).

262. Willow Nestmaker. *Ichthyura apicalis.* ADULT: BL 15–18 mm to tips of wings; mouse gray FW with red-brown shading and a conspicuous white mark just before the apex; antennae feathery, more broadly so in male. LARVA: Dull yellow-brown with 3 faint, dark stripes on the back and with bright yellow-brown protuberances; solitary in nests on willow and poplar. RANGE: Coastal areas, Coast Ranges, and up to moderately high elevations on both sides of the Sierra Nevada.

263. American pussmoths. *Cerura.* Several forms occur in California. ADULTS: BL 18–23 mm to tips of wings; heavy-bodied; white to gray, FW with variable development of broad, blackish cross-bands and dots. LARVAE: Form bizarre with enlarged anterior portion and the anal legs modified into 2 elongate, thin tails; head red-brown, body yellow or greenish with a brown "saddle" area along the back, faintly dotted with brown; on willow or poplar. RANGE: Widespread in arid parts of the state, the Coast Ranges, and Sierra Nevada foothills.

MILLERS AND CUTWORMS
Family Noctuidae

The Noctuidae has more species than any other family of Lepidoptera. Noctuids make up about one-third of all described moths in North America, and probably over 1,000 species occur in California. The adults are stout-bodied with moderately narrow forewings and fan-shaped hindwings which in most species are hidden under the forewings when folded.

Most are cryptically colored, strong-flying, nocturnal moths, but a few are brightly colored dayfliers, often those that are active in early spring or at timberline and in the deserts.

The thick-bodied caterpillars (many of which are called "cutworms") possess all 4 pairs of false body legs. In some species the first 2 pairs are reduced, causing the larvae to walk like geometrid caterpillars, and these noctuids are called "semi-loopers." Cutworms are mostly hairless and cryptically colored, living in leaf shelters or in debris on the ground, from which they emerge to feed at night.

264. Greasy Cutworm or **Black Cutworm.** *Agrotis ypsilon.* ADULT: Medium-sized (BL 22–28 mm to tips of wings); FW brownish clouded with black, with a poorly defined, pale band preceding the terminal edge; HW shining, pearly white with dark veins and edges. LARVA: Greasy looking, gray to brown with a paler stripe down the back; skin covered with large and small, convex, rounded granules; feeding on a wide variety of low-growing succulent plants, including many field crops. RANGE: Widely distributed at low to moderate elevations, from Mt. Shasta to the Colorado River.

265. Variegated Cutworm. *Peridroma saucia.* ADULT: Usually somewhat larger than the preceding species (BL 24–30 mm to tips of wings); FW variably colored, dull reddish brown, gray or blackish brown, sometimes with a broad, poorly defined tan area on the inner half or on the costal margin. LARVA: Variable in color, gray or brown with darker mottling on the sides, a yellow dot on the midline of the back on most segments; on a wide variety of garden plants, truck crops, and trees. RANGE: Low to moderate elevations throughout the state.

266. Early spring millers. *Xylomania, Xylomyges.* Several species of these genera fly in winter and early spring. From December to March they are often seen at lights, even on nights with temperatures near freezing. (**266**) *Xylomania crucialis* is one of the commonest. ADULTS: Medium-sized (BL 18–24 mm to tips of wings); FW whitish gray marked by distinct, longitudinal, black lines and a pinkish tinge in the pale

Figs. 264–274 Miller Moths: 264, Black Cutworm, *Agrotis ypsilon;* 265, Variegated Cutworm, *Peridroma saucia;* 266, Early Spring Miller, *Xylomania crucialis;* 267, Armyworm, *Pseudaletia unipuncta;* 268, Beet Armyworm, *Spodoptera exigua;* 269, February Miller, *Feralia februalis;* 270, Garden Miller, *Galgula partita;* 271, Corn Earworm, *Heliothis zea;* 272, Arizona Desert Miller, *Conochares arizonae;* 273, White Annaphila, *Annaphila diva;* 274, Meadow Miller, *Axenus arvalis.*

areas and on the body. LARVAE: Head bright red; body light gray with smooth skin and faint, broken, pale lines along back and sides; on alder, oak, and other broadleafed trees. RANGE: Coast Ranges, Sierran foothills, and mountains of southern California. Adults from February to May.

267. Armyworm. *Pseudaletia unipuncta.* This is perhaps the most widespread and common noctuid in California; generations occur throughout the year in agricultural and weedy situations. ADULT: Medium-sized (BL 20–26 mm to tips of wings); FW smooth, slightly pointed, nearly uniform pale or dark tan, or speckled with dark dots and mottled with pale tan, a small, whitish mark near the middle and a blackish line leading to the apex; HW gray. LARVA: Smooth, greenish brown to dark gray with 3 yellowish lines down the back and a broader, darker, yellow stripe on each side; usually on grasses but sometimes on field crops, often in large numbers. RANGE: Throughout California west of the Sierra to about 1500 m (5,000 ft) and in the desert mountain ranges.

268. Beet Armyworm. *Spodoptera exigua.* One of the commonest millers in California, occurring in weedy areas and agricultural fields throughout the year. ADULT: Medium-sized (BL 12–19 mm to tips of wings); FW narrow, pale brownish, mottled with gray, having on the outer, leading half, 2 fairly well-defined, whitish spots sometimes containing orange; HW white. LARVA: Pale or yellow-green, a dark line down the back and a broad, dark stripe on each side; on a wide variety of field crops and low, weedy vegetation. RANGE: At low elevations throughout most of the state, including desert areas, foothills up to 1000 m (3500 ft).

Plate 10e. Yellowstriped Armyworm. *Spodoptera ornithogalli.* ADULT: BL 16–26 mm to tips of wings; FW brown with pale lines, discrete tan markings, and a pearly white, blurred streak inward from the wingtip; HW white, narrowly edged in brownish, without a spot underneath. LARVA (**Plate 10e**): Velvety black with 2 prominent and many fine, bright yellow stripes along the sides; general feeder, at times seriously damaging young cotton and other field crops. RANGE: San

Joaquin Valley and Santa Barbara southward. The closely similar **Western Yellowstriped Armyworm** *(S. praefica)* is more widespread in California and is common throughout the Central Valley. It has a paler FW, lacking the white apical streak; HW gray or white with a small, brown spot in the middle of the underside.

269. February Miller. *Feralia februalis.* This species appears at lights with the other harbingers of spring in late January or early February, often on cold nights. ADULT: BL 16–22 mm to tips of wings; strikingly colored, greenish mottled with black and white, apparently cryptically colored resembling lichens; there is a rare form with orange-tan replacing the green; HW white. LARVA: Greenish with yellow flecks and a white midline down the back; eighth abdominal segment with a conical hump and 2 U-shaped marks; on oak, mountain mahogany *(Cercocarpus),* elderberry *(Sambucus),* and probably other plants. RANGE: Coastal areas, coast mountains, and foothills of the Sierra Nevada.

270. Garden Miller. *Galgula partita.* The moths are often seen fluttering about gardens or in weedy areas late in the afternoon; also sometimes attracted to lights at night. ADULT: Rather small (BL 10–14 mm to tips of wings); outline triangular. There are two forms: either **(270)** FW reddish brown, with a dark spot on the middle of the leading margin and a transverse line before the terminal margin, or reddish black with the markings obscured. The blackish form is less numerous but more often noticed due to its unusual appearance. LARVA: Strangely formed with third to fifth body segments swollen into a large hump; green with white stripes along the sides, purplish brown stripes down the back, darker on either side of a white stripe across the hump; on weedy clovers and perhaps other low-growing plants. RANGE: Low elevations along the coast and in the Central Valley, especially in cities.

271. Corn Earworm or **Bollworm.** *Heliothis zea.* This is the most destructive moth and perhaps the most destructive insect to man's economy in California, especially because of its damage to cotton. It is estimated that American farmers grow an average of 2 million acres of corn each year to feed the corn earworm. ADULT: BL 19–25 mm to tips of wings; variable in size and color, FW tan or olive tan, nearly unicolorous or marked with darker bands; HW whitish with a broad, dark band on the outer half; females are more strongly marked. LARVA: Light green, pinkish, or brown with alternating dark and light stripes down the body; in corn ears, cotton bolls, green tomatoes, and other field crops. RANGE: Coastal areas, Central Valley, and deserts.

272. Arizona Desert Miller. *Conochares arizonae.* ADULT: Small (BL 9–12 mm to tips of wings); white with pale brownish pattern on the FW. Common throughout the deserts, often coming to lights in large numbers. LARVA: Apparently unknown. RANGE: Arid parts of the state, north into the San Joaquin Valley.

273. Annaphilas. *Annaphila.* Members of this genus are diurnal, flying in spring; about 10 species occur in California. Their brightly colored hindwings and undersides, red, orange, or white, make annaphilas popular with collectors. **(273) White Annaphila** *(A. diva)* is the commonest species in the Coast Ranges. ADULT: Small (BL 10–20 mm to tips of wings); FW triangulate, black mottled with white; HW white with a black border. LARVA: Head amber colored, body brownish gray with indistinct darker bands down the back and sides, joined by a series of blackish spots on the abdominal segments; each side has a distinct white stripe; reared in captivity on miner's lettuce *(Montia).* RANGE: Coast Ranges and foothills of the Sierra Nevada, to about 1500 m (5,000 ft). Two species with orange HW also fly in the Coast Ranges and foothills of the Sierra Nevada in spring: *A. depicta* has gray FW with a darker, dull reddish brown, transverse band in the outer one-third; wing expanse 21–24 mm; *A. decia* is smaller (wing expanse 18–21 mm) and has a transverse band of white dusting across the FW.

274. Meadow Miller. *Axenus arvalis.* ADULT: Small (BL 9–12 mm to tips of wings); dayflying; a brown moth with yellowish, indistinct bands across the wings. LARVA: Not described. RANGE: Coast Ranges and Sierra Nevada up to 1000 m (3500 ft). Common in spring when wildflowers are in bloom.

275; Plate 10f. Alfalfa Semilooper. *Autographa californica.* One of California's commonest moths, this species is seen throughout the year visiting flowers during the daytime or at lights at night. ADULT (**275**): BL 20–25 mm to tips of wings; gray mottled with blackish and having a silvery, comma-shaped mark near the middle of the FW; HW brownish with a broad, blackish margin. Newly emerged individuals have a curious form when at rest because of upraised scale tufts on the back. LARVA: Olive green with a darker line down the back and a pale stripe along each side; first 2 pairs of false body legs greatly reduced, so the caterpillar walks like a looper; feeds on a wide range of field crops and weedy plants. RANGE: Throughout most of the state at a wide range of elevations. A related species, (**Plate 10f**) **Bilobed Semilooper** *(A. biloba),* is brown with a broad, silver spot; it is less widespread but is common in cities.

276; Plate 10g. Moon Umber. *Zale lunata.* ADULT (**276**): wing expanse 36–50 mm; bark-colored, brown mottled with variable blackish and ochreous. This moth is nocturnal and perches during the day on tree trunks or rocks with the wings spread out flat, like a geometrid moth. LARVA (**Plate 10g**): Gray to brown mottled with blackish, with roughened skin and large, conical tubercles on the eighth abdominal segment; with projecting flaps that lie against the substrate, resulting in strong cryptic resemblance to tree branches on which they rest; feeding on blackberries *(Rubus),* oaks, willows, and other plants. RANGE: Coastal Plain, foothills, and Central Valley. Often seen at lights in cities.

277. Black Witch. *Otosema odorata* (formerly *Erebus odora*). As big as a bat, this species has the largest wingspread (12–17 cm) of any insect in California, and its appearance at lights around a home or store never fails to attract notice. Individuals at times fly far north of the normal range, startling

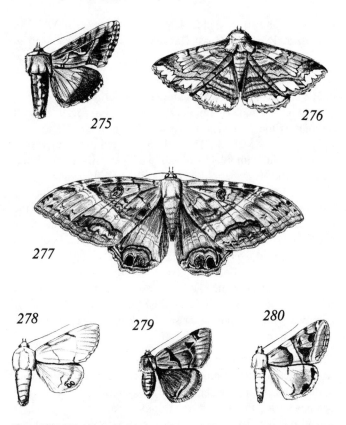

Figs. 275–280 Noctuid Moths: 275, Alfalfa Semilooper, *Autographa californica;* 276, Moon Umber, *Zale lunata;* 277, Black Witch, *Otosema odorata;* 278, Palm Moth, *Litoprosopus coachella;* 279, Duskywing Moth, *Euclidina ardita;* 280, Desert False Underwing, *Drasteria tejonica.*

residents in places like San Francisco or Reno. ADULT: FW triangular; HW broad, rounded; dark brown or blackish with blackish lines and circular eyespots tinged with blue; female with a jagged white band across the wings. LARVA: Up to 7 cm long; tapered posteriorly; dark gray tinged with brown, with a pale stripe down the back and dark stripes down the sides; feeds on acacia, cassia, and other plants of the pea family. RANGE: A native of Mexico, possibly resident in San Diego, breeding on ornamental plants; commonly migrating northward as far as Los Angeles, from July through the fall months.

278. Palm Moth. *Litoprosopus coachella.* Caterpillars of this species normally live in palm trees but sometimes invade buildings in search of cocoon sites. They have been known to burrow into stored fabrics, paper products, and acoustic tiling. ADULT: BL 20–26 mm to tips of wings; shining cream colored or pale tan; FW with faint, darker crosslines and twin, black 00's on the margin of HW. LARVA: Polished, smooth-skinned, brownish with a pale stripe down the back that includes dark, longitudinal lines on most segments; on stems, fronds, flowers, and fruits of fan palms *(Washingtonia)*. RANGE: Desert areas and cities of southern California.

279. Duskywing Moth. *Euclidina ardita.* This moth flies in spring during the daytime and is often mistaken for a **(301)** duskywing skipper *(Erynnis)*. There are even accounts in the literature of attempted matings between humbugged male *Erynnis* and these moths. ADULT: BL 17–20 mm to tips of wings; dark brown or blackish brown, FW with black and gray markings; HW with an indistinct, pale brown band. LARVA: Pale greenish with a pale stripe down the back including a dark-bordered, reddish line, and dark stripes along sides just above and including the legs; on clover, lupine, and grasses. RANGE: Coast and southern ranges and Sierra Nevada, foothills up to 1500 m (5,000 ft).

Plate 10h. Underwings. *Catocala.* These large noctuids are among the more showy moths of California, but they are not commonly seen because of their cryptic color and behavior. A method of attracting *Catocala* to sweet, fermenting baits ("sugaring") is popular with collectors in the eastern United States where underwings are much more abundant. It is most successful on warm, humid evenings. Results usually are poor in California. ADULTS: Moderately large (BL 24–44 mm to tips of wings); FW gray, triangulate, mottled, resembling tree bark, where the moths rest by day; HW red, orange, or white with black markings, covered when the wings are folded. When the insect is disturbed it flashes the hindwings suddenly, and this is believed to have a startling effect on potential predators such as birds. LARVAE: Gray, tapering posteriorly, with a fringe of lateral projections; often a prominence on the fifth abdominal segment and raised tubercles, especially on the eighth segment; on broadleafed trees including poplar, willow, oaks, and alders. RANGE: Wooded parts of the state. Attraction of underwings to lights seems to be erratic, and some species are more frequently seen at their resting stations during the daytime. Hunting is sometimes productive if one searches the trunks of trees or under loose bark, of cottonwood, alder, etc. Other hiding places include the vertical surfaces of granite boulders and under eaves or bridges in wooded areas.

280. False underwings. *Drasteria.* Several species resemble true underwings in having the FW cryptically colored and HW orange, marked with black. **(280) Desert False Underwing** *(D. tejonica)* has the FW strongly marked with white; HW mostly orange in the female, white tinged with pink or orange in the male; 32–38 mm wingspread. RANGE: Desert areas and Central Valley north to Antioch; sometimes appearing in great numbers at lights on warm nights. A similar species, **Edwards' False Underwing** *(D. edwardsi),* has the HW orange, heavily marked with black, and a dark pattern narrowly outlined in yellow on FW. It is common in the Coast Ranges and Sierra Nevada at mid-elevations, often seen flying in the daytime.

Family Agaristidae

Agaristids are dayflying moths resembling noctuids and tiger moths, with the antennae slightly enlarged towards the tip like those of sphinx moths. Only 7 species occur in California.

281. Ridings' Forester. *Alypia ridingsi.* ADULT: Wing expanse 30–34 mm; black with white spots divided by black veins; middle pair of legs with bright orange scaling. LARVA: Brightly colored with raised, black bands around the segments and orange or yellow spots; humped toward the rear; feeds on farewell-to-spring *(Clarkia)* in late May. RANGE: Coast Ranges, Sierra Nevada, and desert ranges, occupying widely diverse habitats from near the coast to above 2500 m (8,000 ft). *A. mariposa* is very similar but lacks the black

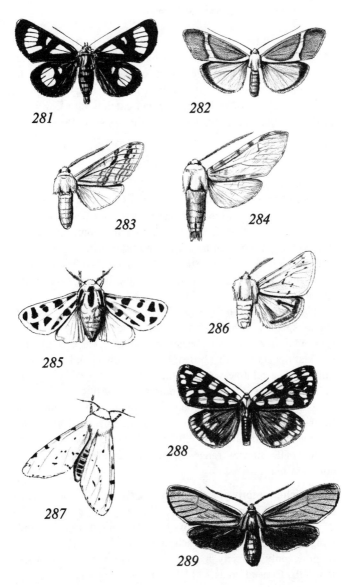

Figs. 281–289 Moths: 281, Ridings' Forester, *Alypia ridingsi;* 282, Lichen Moth, *Cisthene* species; 283, Yellow-Spotted Tiger Moth, *Halisidota maculata;* 284, Edwards' Glassywing, *Hemihyalea edwardsi;* 285, Mexican Tiger Moth, *Apantesis proxima;* 286, Wandering Tiger Moth, *Spilosoma vagans;* 287, Acrea Moth, *Estigmene acrea;* 288, Ranchman's Tiger Moth, *Platyprepia guttata;* 289, Brown Ctenucha, *Ctenucha brunnea.*

veins through the wing markings; the **Eight-spotted Forester**
(A. octomaculata) differs by having orange on both the fore-
and mid-legs and by having 2 spots on each wing, those of the
FW yellowish, those of the HW white. It is a widespread
species in eastern North America and extends into California
in the northern counties and along the coast as far south as
Santa Maria.

TIGER MOTHS
Family Arctiidae

Arctiids are stout-bodied, furry, and often brightly colored
moths, usually with the antennae narrow and feathery in the
male, slender and filamentlike in the female. Some species are
dayflying, but most are nocturnal and often attracted to lights.
The caterpillars characteristically are covered with long, erect
hairs and are popularly called "woollybears." About 50
species are known in California.

The **Banded Woollybear** *(Pyrrharctia isabella),* a widespread species in the eastern United States, is alleged to possess predictive powers for weather forecasts. The caterpillar is
black at the ends, with a broad, reddish brown band around the
middle. The relative width of the colors is supposed to be an
index to the severity of the coming winter. Although this may
be as good a method for long-range predictions as any used
by the weather bureau, if the colors are correlated with temperature or rainfall, probably they indicate past not future
events. The species ranges west into California but is not
common here.

282. Lichen moths or **footmen.** *Cisthene.* Several species
occur in California. ADULTS: Small (wing expanse 17–
25 mm); FW gray, with a cross-band of white or like the hindwing, pink, salmon, pale peach, or yellow. LARVAE: Blackish
or gray, only sparsely hairy; on lichens on rocks and bark.
RANGE: Coastal areas and Coast Range foothills.

**283; cover photo (upper right). Yellow-spotted Tiger
Moth.** *Halisidota maculata.* ADULT (**283**): BL 20–30 mm to
tips of wings; FW yellow, variably marked with brownish; HW
whitish. Common at lights in spring. LARVA (**cover photo**):

Densely hairy, black with intermixed white hairs and a wide, yellow band around the middle; on willow and other broad-leafed trees and shrubs. RANGE: Coastal areas and mountains and both sides of the Sierra Nevada to about 2000 m (6500 ft). A related species, the **Silver-spotted Tiger Moth** *(H. argentata),* has brown FW marked with silver-white spots. Its larva feeds on coniferous trees in the Coast Ranges, Sierra Nevada, and southern mountains.

284. Edwards' Glassywing. *Hemihylaea edwardsi.* ADULT: One of the state's largest tiger moths (BL 30–38 mm to tips of wings), this species is unusual in having the wings sparsely scaled and translucent. Wings tan, dusted with blackish, obscure transverse bands in fresh specimens; abdomen bright rose-pink. LARVA: Head large, shiny dark brown; body velvety brownish black, densely clothed with tufts of brown hairs; underside and legs yellowish; on oaks and possibly other plants. RANGE: Coastal areas and mountains and foothills of the Sierra Nevada.

285. Mexican Tiger Moth. *Apantesis proxima.* This is the commonest tiger moth in southern California. ADULT: Male **(285)** has black and white checkered FW; white HW; and rose-pink abdomen. Female has yellow and black FW; abdomen and HW rose-red spotted with black; BL 20–28 mm to tips of wings. LARVA: Densely hairy, black; on mallow and various low-growing plants, often found under debris in fall and winter. RANGE: Arid parts of the state north into the Central Valley. A similar species, the **Ornate Tiger Moth** *(A. ornata),* is less numerous in southern California but males are common at lights in the central and northern Coast Ranges in spring. FW with pale areas tinged pink or peach colored; there are 2 forms, one with yellow, one with red HW. The amount of black spotting is variable and more extensive in the north. Females are rare in collections and may be diurnal.

286. Wandering Tiger Moth. *Spilosoma vagans.* ADULT: Male pale tan to gray-brown; female reddish brown; FW with variable blackish lines; HW dark gray with a border of FW color; BL 20–24 mm to tips of wings. LARVA: Brownish black with reddish legs, red-brown hair, and some longer, black hairs posteriorly; on various low-growing herbs. RANGE: Coastal areas and mountains and Sierra Nevada up to about 2000 m (6500 ft). The females apparently do not come to lights and may be dayflying.

Plate 11a. Vestal Tiger Moth. *Spilosoma vestalis.* Flying only in spring, this moth is easily recognized by its bright red front legs. ADULT: BL 22–32 mm to tips of wings; white with black dots on the wings and black bands on the abdomen. LARVA: Black, densely clothed with relatively short, stiff hairs; on various low-growing plants. RANGE: Coastal areas and mountains to about 1500 m (5,000 ft).

287. Saltmarsh Caterpillar or **Acrea Moth.** *Estigmene acrea.* Probably the most widespread and often seen tiger moth in California, this species is sometimes abundant enough in field crops to be of economic importance. ADULT: BL 24–34 mm to tips of wings; female has white wings spotted with black and a yellow abdomen; male similar but has ochreous yellow HW. LARVA: Black with yellow, broken lines, covered with clumps of long hairs and red-brown hairs at the sides. Most abundant in fall; on various low-growing plants. RANGE: Coastal areas, Central Valley, and deserts.

Plate 11b, 11c. Painted Tiger Moth. *Arachnis picta.* ADULT (**Plate 11b**): BL 22–35 mm to tips of wings; FW gray marked with wavy, whitish lines, giving the insect a lichenlike appearance when at rest; FW underneath yellow on the basal half; HW and abdomen bright pink with gray markings. Common in fall at lights. LARVA (**Plate 11c**): Densely clothed with long, stiff, black and brown hairs; on various herbs and shrubs in winter and spring. RANGE: Coastal and arid parts of the state, in the Central Valley north to Antioch. Populations in the mountains of the eastern Mojave and Inyo County consist of larger moths with restricted gray markings and bright rose-pink hindwings.

288. Rangeland or **Ranchman's Tiger Moth.** *Platyprepia guttata.* California's largest arctiid. ADULT: BL 34–40 mm to tips of wings; antennae threadlike in both sexes; wings exceptionally broad; FW blackish blue with cream-colored spots; HW and abdomen yellow with black spots, or almost entirely black. LARVA: Densely covered with elongate, red-brown, white, and black hairs; on bush lupine and probably other plants. RANGE: East of the Sierra Nevada and localized spots in other parts of northern California; common at Pt. Reyes National Seashore.

Family Amatidae (Ctenuchidae)

The family Amatidae is primarily tropical and consists mainly of brightly colored, dayflying moths that resemble wasps or beetles which are distasteful to vertebrate predators. Amatids differ from tiger moths only in minor structural features of the wing venation. Larvae and pupae of the 2 families are similar. Fewer than 10 species occur in California.

289. Brown Ctenucha. *Ctenucha brunnea.* ADULT: BL 20–26 mm to tips of wings; body metallic blue with pale reddish shoulder straps extended back over the thorax; FW brownish with the leading edge white, veins blackish; HW black. LARVA: Head orange, body blackish with longitudinal yellow stripes; long, erect, whitish or tan hairs on the back with black tufts on the fourth and next to last body segments; feeds on sedges and grasses. Pupation occurs in a soft cocoon of felted hairs shed by the larva. RANGE: Coastal areas of southern California. North of Santa Maria the brown disappears from the FW and the name *C. multifaria* is used. A related species, the **Red-shouldered Ctenucha** *(C. rubroscapus),* occurs along the north coast and inland in the Coast and Cascade ranges and the Sierra Nevada; FW black without the white leading edge.

SILK MOTHS
Family Saturniidae

The family Saturniidae includes some of the largest and most beautiful Lepidoptera. They are broad-winged moths with stout, short bodies and feathery antennae in both sexes, broader in males. The caterpillars are not hairy but are usually armed with protuberances or many-branched spines, which in some species cause a skin rash like that caused by nettle. Pupation occurs underground or in a dense, silken cocoon. In the Orient some species yield commercial silk, but the principal Silk Moth *(Bombyx mori)* is an entirely domestic animal. It is a member of the Bombycidae, a family which has no native species in California. Saturniids are primarily tropical, and California, with only about 15 species, has a poor representation compared to Mexico or the eastern United States. There are 2 subfamilies in our fauna: the Saturniinae are mostly large, nocturnal moths with spineless larvae; and the Hemileucinae are smaller, often dayflying moths, whose caterpillars have branched spines.

Plate 11d, 11e. Ceanothus Silk Moth. *Hyalophora euryalus.* This is probably California's most spectacular and commonly seen large moth. Females release a chemical during early morning hours which attracts males. The moths are most active and often come to lights between midnight and dawn and are seen clinging to porches or store fronts in the morning, especially in the spring. ADULT (**Plate 11d**): Broad-winged (8–12.5 cm across wings when spread); with beautiful rose red uppersides marked by white, crescent-shaped spots and cross-lines of white; underside grayish, white-flecked, and tinged with reddish. LARVA (**Plate 11e**): Pale apple green with yellow protuberances on top of each segment, those of the first 3 segments larger and banded with black; with smaller, pale bluish tubercles on the sides; feed during late spring and summer, reaching a length of 10 cm; on *Ceanothus,* coffeeberry *(Rhamnus),* willow, and other shrubs. COCOON: Gray, oval with the emergence end drawn out; on branches of the host plant. RANGE: Coastal areas and mountains up to 2700 m (9,000 ft) elevation.

290. Polyphemus Moth. *Antheraea polyphemus.* This species is slightly larger than the preceding (wing expanse 10–14 cm) but is not as commonly seen in most parts of California. ADULT: The wings usually are tan (rarely tinged with olivaceous or reddish brown), each wing with a scaleless, windowlike, yellow-margined, circular spot; those on HW surrounded by a large color spot of blue and black, resembling an eye; when the insect opens its wings upon disturbance, the illusion is that of eyes suddenly opening, and this is believed to frighten would-be predators. LARVA: Pale green with oblique, pale yellow lines on each side and short tubercles arising from red spots; feeds on a wide variety of broadleafed trees and shrubs; occasionally a pest in stonefruit orchards. COCOON: Whitish, oval without a drawn-out emergence neck. RANGE: Throughout most of the state, foothills to moderate elevations.

291. Pandora Moth. *Coloradia pandora.* This species is reported to have a two-year life cycle, overwintering first as a young larva and spending the second winter as a pupa, with adults of a given population appearing in alternate years. ADULT: Wing expanse 7–9 cm; heavy-bodied; drab brownish gray with black lines and a single black spot on each wing; HW paler, tinged with pinkish. LARVA: Green or greenish brown with 6 broad but broken, interrupted, longitudinal dark stripes, with short, spiny tubercles; feeds on pines. RANGE: Western Yellow Pine and Jeffrey Pine areas in southern California and

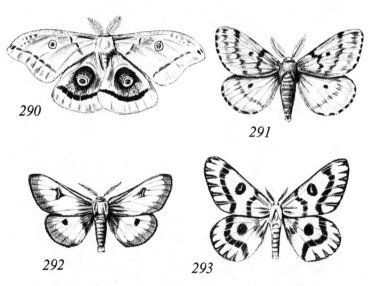

Figs. 290–293 Silk Moths: 290, Polyphemus Moth, *Antheraea polyphemus;* 291, Pandora Moth, *Coloradia pandora;* 292, Nevada Buck Moth, *Hemileuca nevadensis;* 293, Sheep Moth, *H. eglanterina.*

east of the Sierra Nevada. As late as the early 1900s, Pandora Moth larvae and pupae were eaten by Indians in the Mono and Modoc areas, who dug trenches beneath the trees to trap larvae as they crawled down, and then baked the caterpillars in hot soil and dried them for later consumption.

292. Nevada Buck Moth. *Hemileuca nevadensis.* ADULT: Wing expanse 5–7.5 cm; white or yellowish with a dark abdomen having an orange or reddish tip; wings blackish at base and apical margin, with a dark blotch surrounding a white spot on FW. LARVA: Black with yellowish stripes down back and sides; with yellowish or brownish, branched spines that render a stinging rash to sensitive areas of human skin; feeding on willow and poplar. RANGE: Coastal Plain in southern California, Central Valley, and east of the Sierra Nevada. A related species, the **Electra Buck Moth** *(H. electra),* has bright red-orange HW bordered with black; its larvae feed on buckwheat *(Eriogonum)* in southern California. Both species fly during the daytime in the fall. The common name derives from a widespread eastern species which also flies in fall, during deer-hunting season. The eggs of *Hemileuca* are deposited in a mass encircling a branch or twig and are easily seen on deciduous plants in winter.

293. Common Sheep Moth. *Hemileuca (Pseudohazis) eglanterina.* This dayflying moth appears in summer. ADULT: Wing expanse 5.5–8.5 cm; relatively slender-bodied; FW typically pink with a yellow streak in the middle, outer area; HW yellowish; variable black markings; an almost entirely black form occurs near Mt. Shasta, and rarely individuals lack nearly all the black. LARVA: Black when young, with many-branched spines; head orange-brown and spines yellowish or orange when full grown; not as readily stinging to the human skin as are buck moths; on wild rose, *Ceanothus,* coffeeberry *(Rhamnus),* and other shrubs. RANGE: Throughout much of the state west of the Sierran crest; absent from San Joaquin Valley except in the Delta; restricted in southern California to the mountains. Two similar species occur in sagebrush areas east of the Sierra Nevada: **Nuttall's Sheep Moth** *(H. nuttalli),* the larva of which feeds on snowberry *(Symphoricarpos);*

and the **Hera Moth** *(H. hera),* which feeds on sagebrush *(Artemisia).* Both are pale colored, lacking the pink tinge of the FW; *H. hera* has whitish wings, *H. nuttalli* has yellow HW.

HAWK MOTHS, SPHINX MOTHS
Family Sphingidae

Sphingids are moderate-sized to large moths with streamlined, stout bodies and elongate forewings having a strongly oblique outer margin. The antennae are thickened towards the middle and tapered towards the tip, which is usually slightly hooked. Many species have a very long tongue, 2 or 3 times the length of the body, which is used to feed in deep-throated flowers. Hawk moths are important pollinators for some plants. Most sphingids are nocturnal, but some genera are dayflying moths that resemble hummingbirds or large bees when visiting flowers.

The caterpillars are smooth skinned, without hairs or spines, and the eighth abdominal segment nearly always bears a thin spur or horn on the back. A wide variety of plants is eaten, although many species are quite specific in host selection, with larvae cryptically colored like the plant. The family name stems from the alarm posture of the larva. With its forequarter reared back and head tucked downward, it resembles the form of the Egyptian sphinx. Pupation occurs in a cell in the ground, without a cocoon.

Most of the world's hawk moths occur in tropical regions. In California we have only about 25 species, a small number compared to the southeastern United States or Mexico.

294; Plate 11f. Eyed Sphinx. *Smerinthus cerisyi.* This species is one of California's most common sphingids and is often attracted to lights, especially late at night. ADULT (**294; Plate 11f**): Moderate-sized (BL 20–32 mm; wing expanse to 9 cm); FW ragged-margined, brownish gray or tan, with pale markings; HW rose pink, with dark blue circle enclosing a whitish spot, resembling an eye. When perched, the insect has a cryptic appearance, like dead leaves, but when disturbed the wings are brought forward, exposing the eyespots of the HW (**Plate 11f**). The sudden appearance of the eyespots is believed

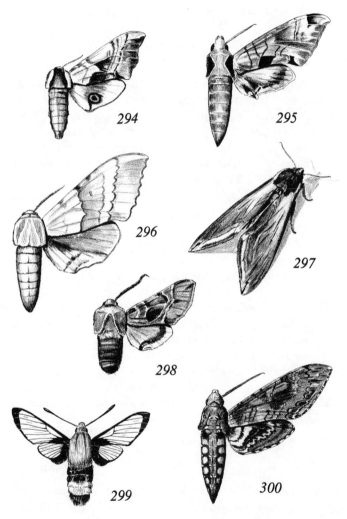

Figs. 294–300 Sphinx Moths: 294, Eyed Sphinx, *Smerinthus cerisyi;* 295, Achemon Sphinx, *Eumorpha achemon;* 296, Western Poplar Sphinx, *Pachysphinx occidentalis;* 297, Elegant Sphinx, *Sphinx perelegans;* 298, Bear Sphinx, *Arctonotus lucidus;* 299, Bumble Bee Moth, *Hemaris diffinis;* 300, Tomato Hornworm, *Manduca sexta.*

to frighten potential predators. LARVA: Green with a pair of subdorsal, pale yellow lines and 6 pairs of oblique, white bands above the spiracles; horn yellow, pink, or blue; feeds on willow and poplar. RANGE: Throughout most of the state at low elevations.

295. Achemon Sphinx. *Eumorpha achemon.* ADULT: BL 36–48 mm, wingspread to 11.5 cm; body and FW brownish gray with black and dark brown markings; HW rose pink on basal half; underside reddish. The moths are nocturnal, flying in late spring and again in midsummer. LARVA: Green or reddish brown, coarsely granulated with pale specks, with a whitish line along the top of each side, below which are 7 or 8 large, oval, oblique, white spots; anal horn long and thin in early stages, but is normally lost before maturity and replaced by a raised spot; on grapevines and Virginia Creeper. RANGE: Widespread in the state at low elevations.

296. Western Poplar Sphinx. *Pachysphinx occidentalis.* This is the biggest sphingid in California. ADULT: BL 38–50 mm; wing expanse to 15 cm. There are 2 color forms, one pale tan with faint brownish crossbanding on the FW, and the HW pale rose pink with an indistinct pale bluish eyespot; the other form is darker, gray-brown with a rose-purplish HW. LARVA: Green with fine, white specks, pale, oblique, lateral stripes, and a long, brown anal horn; on poplar and occasionally willow. RANGE: Low elevations throughout the state.

297. Elegant Sphinx. *Sphinx perelegans.* ADULT: BL 34–44 mm; wing expanse to 10.5 cm; FW nearly uniform gray, with a white band before the terminal margin; HW and abdomen banded black and white, thorax entirely black above. LARVA: Bright green with broad, oblique bands on each side and a brown horn; on manzanita *(Arctostaphylos)* and shrubs of the rose family. RANGE: Coastal areas and mountains and Sierra Nevada up to about 2,000 m (6500 ft). Several similar-appearing species occur in California, including the **Chersis Sphinx** *(S. chersis)* which is widespread but less common. It is larger (wing expanse 9–11.5 cm), is paler gray, and has the thorax pale gray with 2 black lines the length of the dorsum.

The **Sequoia Sphinx** *(S. sequoiae)* is similarly colored but is much smaller (wing expanse 5.5–6 cm).

298. Bear Sphinx. *Arctonotus lucidus.* In January and February this small, fat, green sphingid appears at lights, often on rainy or unusually cold nights when little else is flying. ADULT: BL 18–24 mm; wingspread 40–50 mm; thickly clothed with woolly-appearing scaling; body and FW dull olive green with purplish bands across the FW; HW dull reddish with a greenish border. LARVA: Greenish black, venter gray-green; dorsum with transverse white bands, each with a black patch in the middle; segments 4–9 with pairs of white dots and a row of oblique, white lines below the spiracles; reared in captivity on evening primrose *(Oenothera)*. RANGE: Foothills of Coast Ranges, Sierra Nevada, and southern mountains.

299. Bumble Bee Moth or **Snowberry Clearwing Sphinx.** *Hemaris diffinis.* This moth flies during the daytime in late spring and summer and resembles a bumble bee as it hovers at flowers for nectar. ADULT: BL 16–23 mm; wings scaleless except reddish black at base, on the veins, and at the outer margin; body clothed with fluffy scaling, thorax olive to golden brown, abdomen black with a broad, yellow band. LARVA: Length to 45 mm; variable in color, pale green, thickly granulated with white, reddish beneath; thoracic shield with a raised, double row of yellow granules; body with a pale, lateral stripe; horn long, slender but sometimes lost; on dogbane *(Apocynum)* and more commonly snowberry *(Symphoricarpos)* and other members of the honeysuckle family. RANGE: Coastal valleys and mountains throughout the state, often to 2,000 m (6500 ft), rarely to 2,700 m (9,000 ft).

Plate 11g, 11h. White-lined Sphinx. *Hyles lineata.* This is the most common and widespread sphingid in California, and was formerly known as *Celerio lineata.* In late afternoon and at dusk the moths are often seen hovering like hummingbirds at petunias and other garden plants, and later at night they come to lights. ADULT (**Plate 11g**): Moderate-sized (BL 25–45 mm; wingspread to 9.5 cm); body and FW brown with pale tan markings; abdomen spotted with black and white; HW pale rose pink with basal and marginal black bands. LARVA (**Plate**

11h): Variable in color, ranging from green with thin, black lines to nearly entirely black with thin, pale lines; with the head and the short horn yellow or orange; on a wide variety of low-growing herbaceous plants. RANGE: Throughout the state. Occasionally occurs in tremendous outbreaks, especially in desert areas. The larvae migrate in great hordes, devouring all annual vegetation, and have even been reported to slicken highways with their crushed bodies.

300; Plate 12a. Tomato Hornworm. *Manduca sexta.* This species is one of California's largest and commonest sphingids. Its huge green caterpillar is familiar to home gardeners who grow tomatoes. ADULT (**300**): BL 40–55 mm; wing expanse 10–12 cm; FW dark gray with whitish and blackish wavy streaks and a distinct, white dot in the middle towards the leading edge; HW gray with indistinct white bands; abdomen with 6 yellow spots along each side. LARVA (**Plate 12a**): BL to 10 cm; green, blending well with tomato leaves, with oblique, white streaks across the sides of body segments; tail horn brown or reddish; on tomato and other plants of the nightshade family. PUPA: Dark brown with the tongue sheath separated, like a pitcher handle. There are 2 generations each year. RANGE: Central Valley and southward at low to moderate elevations. The similar **Five-spotted Hawkmoth** *(M. quinquemaculata)* is less common in California but nearly as widespread. It is often larger (wing expanse 11–13 cm) and has wavy black lines on the FW but no white dot, more distinct, wavy black lines on the HW and usually 5 yellow spots on each side of the abdomen. *M. sexta* was long known in popular literature as the Tomato Hornworm and *M. quinquemaculata* as the Tobacco Hornworm. However, the Entomological Society of America recently reversed the common names in its official list, which dictates usage by most professional entomologists, producing unnecessary confusion that will persist indefinitely.

SKIPPERS
Family Hesperiidae

Compared to true butterflies, skippers are stout-bodied with relatively short wings. The head is large, and the antennae are widely separated at their bases, with the tip prolonged beyond the club to form a small, recurved point. The larvae **(Plate 12b)** are stout with a large head separated from the body by a constricted neck. They make silken shelters, and a weak cocoon is formed for pupation, in contrast to most butterflies.

There are 2 subfamilies represented in California, the Pyrginae, which are mostly black and white skippers whose larvae feed on a wide variety of broadleafed plants, and the Hesperiinae, which are typically orange and brown species that feed on grasses. The family is a large and diverse one with more than 50 species in California. Representatives occur in virtually every habitat, from seacoast and low deserts to alpine places above timberline.

The common name "skipper" derives from the powerful, erratic, jerky flight, which differs from the more uniform, fluttering or gliding maneuvers of true butterflies. When perched, skippers hold the forewings upright and the hindwings spread laterally, unlike true butterflies.

301; Plate 12b. Funereal Duskywing. *Erynnis funeralis.* This is the most widely distributed species of its genus in California among several which appear similar. ADULT (**301**): Wing expanse 35–45 mm; blackish brown; FW relatively narrow with an indistinct pale area on the outer, leading half; HW with white marginal fringes. LARVA (**Plate 12b**): Head black with yellowish spots; body pale green, yellowish in the folds, marked with a yellow line on each side, and bearing a bright yellow spot on each segment; skin covered by short, pale yellowish hairs; feeds on various plants of the pea family. RANGE: Low to moderate elevations from the Central Valley southward through southern California. There are a number of generations each season and adults may be seen from February to October. The **Propertius Duskywing** *(E. propertius)* is more common in central and northern California in the foothills to middle elevations. It is brown, with broader FW than in the preceding species, and the HW lacks the white fringe; the larva feeds on oaks.

Plate 12c. Common Checkered Skipper. *Pyrgus communis.* This is one of the most often seen skippers in California, occurring in a wide range of habitats throughout the year. ADULT:

Medium-sized (wing expanse 28–35 mm); dark gray or black-ish, checkered with white, body and basal areas of the wings covered by white hairs. LARVA: Yellowish white with deli-cate, gray-green lines along the middle and sides of the back; head and collar black; on various plants of the mallow family.

302. Fiery Skipper. *Hylephila phyleus.* This is the most abundant skipper around lawns in California's cities. ADULT: Males are bright yellow-orange, with blackish markings; females brown with yellow-orange markings; easily recog-nized by the yellow undersides speckled with small, brownish dots; wing expanse 25–37 mm. LARVA: Yellowish brown to gray-brown with a black line along the back; head black with pale lines on the face; on various grasses, including lawn varieties. RANGE: Urban and other lowland areas from San Francisco Bay area southward into the deserts. Abundant in southern California, where adults may be seen from March to December.

301

302

303

304

Figs. 301–304 Skippers: 301, Funereal Duskywing, *Erynnis funeralis;* 302, Fiery Skipper, *Hylephila phyleus;* 303, Woodland Skipper, *Ochlodes syl-vanoides;* 304, Stephens' Skipper, *Agathymus stephensi.*

303. Woodland Skipper. *Ochlodes sylvanoides*. This is the commonest skipper throughout the foothills and middle-elevation areas of the state in summer and fall. ADULT: Wing expanse 24–30 mm; deep ochreous orange in both sexes, FW with a strong, black diagonal mark in male; HW underside variable, from uniform ochreous orange to heavily mottled with purplish brown, defining a jagged, median band of ochreous orange. LARVA: Head black; body yellow-tan with 7 blackish, longitudinal lines; on various grasses. RANGE: Coast Ranges and Sierra Nevada to middle elevations and east of the Sierra.

Plate 12d. Sandhill Skipper. *Polites sabuleti*. Common in sandy, natural situations, less prevalent in cities. ADULT: Male orange with dark-margined wings and a black, sigmoid mark on the FW; female with more extensive dark markings but lacks the sigmoid brand; underneath both sexes have characteristic, jagged, pale markings on the HW (**Plate 12d**). Individuals are darker at high elevations, and there is a paler race in the deserts. LARVA: Green with a black head marked by 4 vertical, pale lines on the face; on Salt Grass *(Distichlis spicata)* and other grasses. RANGE: Widespread; especially on seacoast and riverbed dunes but also common at timberline in the Sierra Nevada and the southern mountains.

GIANT SKIPPERS
Family Megathymidae

Adult giant skippers are heavy-bodied insects, larger than most skippers (Hesperiidae), and have clubbed antennae that are not hooked. The larvae have a small head that is partially retractile into the front body segments, and they live inside roots and stems of yucca and other plants of the agave family. A year is required for the life cycle in most species, adults of different species flying in spring, summer, or fall. This is a small family, restricted to arid parts of the United States and Mexico, and only 4 species are recorded in California.

304. Stephens' Skipper. *Agathymus stephensi*. ADULT: Wing expanse 48–58; brown with pale yellowish markings on the upperside; underside with much whitish gray scaling.

LARVA: Stout, whitish with a dark head, neckstrap, and tip of abdomen; burrows in leaves of Desert Agave, pupating in a chamber near the base of the leaf. The adult leaves through a ''trap-door'' constructed by the larva prior to pupation. RANGE: Along the western edge of the Colorado Desert from the vicinity of Palm Springs southward. Megathymids are extremely fast fliers and, even if a collector is fortunate enough to net one, it batters itself to a frazzle before it can be corralled. Thus, specimens are best obtained by gathering nearly full-grown larvae for rearing. Adults of Stephens' Skipper come to wet places to feed at sundown and may be captured by placing a killing jar over them.

Family Lycaenidae

Lycaenids are small butterflies, predominately with metallic blue, coppery, orange, or gray wings. The antennae are ringed with white, and frequently the undersides are whitish or some cryptic color which differs from the uppersides. The wings are held closed above the back when the insect is at rest. The larvae (**Plate 12e**) are grublike, tapering towards the ends, with broad, projecting lobes that conceal the head and legs. Most are clothed with short hairs, their form and color often blending with the plant on which they live. Most lycaenids are specific in foodplant selection. Pupation occurs without a cocoon, and the pupa is stout, rounded in front and smooth with little or no freedom of motion of the abdominal segments.

There are 3 subfamilies in California; the Plebejinae, or **Blues,** of which there are about 25 species; the Lycaeninae, or **Coppers,** represented by 12 species; and the Theclinae, or **Hairstreaks,** with about 30 species.

305. Pigmy Blue. *Brephidium exilis.* The smallest and one of the most common butterflies in California, the Pygmy Blue is especially abundant wherever Australian Saltbush *(Atriplex semibaccata)* grows, flying throughout the year in coastal areas. ADULT: Small (wing expanse 14–22 mm); with rounded wings that are brown or reddish brown above with variable amounts of blue basally and a white marginal fringe; wings underneath pale brownish spotted with white, HW with

Figs. 305–308 Blues: 305, Pigmy Blue, *Brephidium exilis;* 306, Acmon Blue, *Plebejus acmon* (underside); 307, Echo Blue, *Celastrina argiolus;* 308, Xerces Blue, *Glaucopsyche xerces* (underside).

a row of metallic spots along the outer margin. LARVA: Head black, body yellowish green, coarsely granulated with white specks; usually with yellowish lines down the back and sides; feeds on pigweed *(Chenopodium)* and saltbush *(Atriplex)*. RANGE: Low-elevation areas of the San Francisco Bay area and Central Valley and throughout southern California; from alkaline waste places along the coast to Death Valley and up to 2000 m (6500 ft) in desert ranges.

306. Acmon Blue. *Plebejus acmon.* This species is the most generally distributed blue in California. ADULT: Wing expanse 17–30 mm; wings metallic sky blue in the male, brown with variable basal blue in the female; HW with an orange band on the outer margin, pinkish and often reduced to a trace in the male; undersides whitish with black dots and a row of metallic greenish dots at the edge of the orange band, color variable with the seasons. LARVA: Yellowish with a narrow, green stripe down the back and variable markings on the sides, covered with short, white hairs; foodplants are Deerweed *(Lotus scoparius)* and other plants of the pea family and buckwheats *(Eriogonum)*. RANGE: Throughout the state, from sea level up to 3000 m (10,000 ft) in the southern mountains.

307. Echo Blue. *Celastrina argiolus.* This species is widespread but is usually seen only in small numbers, often flying in chaparral-covered areas. ADULT: Wing expanse 22–35 mm; wings metallic powder blue with the leading and outer edges broadly margined with black in the female; darker at high elevations; undersides white with faint, grayish speckling. LARVA: Variable in color and markings, from green with reddish pattern to magenta with pale markings; on various shrubs and trees. RANGE: Throughout the state at a wide range of elevations. Adults fly from February to July in southern California, but the season is shorter at high elevations.

Plate 12f. Sonora Blue. *Philotes sonorensis.* This is one of our most attractive butterflies, but it occurs only in local colonies and flies in early spring, so escapes notice by all but the more industrious collectors. ADULT: Wing expanse 20–26 mm; wings metallic sky blue on the upperside, marked with a bright red-orange spot on the FW in the male and on both wings in the female; black marginal markings also more extensive in the female; undersides gray with whitish and black markings and the orange spots repeated. LARVA: Variable in color, pale green to mottled rose; on Desert Saviour *(Dudleya lanceolata)* and other *Dudleya* species. RANGE: Rocky canyons of the Sierran foothills and Coast Ranges from San Jose and Auburn southward to the desert margins.

Plate 12g. Square-spotted Blue. *Pseudophilotes battoides.* This little blue has diverged into several geographical races with minor differences in size and color. ADULT: Wing expanse 17–27 mm; males are deep blue with a black margin on the wings; females are brown with a broad, orange bar just inside the HW margin; underside whitish with black spots and orange bars on the HW in both sexes. Typical populations occur at timberline in the high Sierra and have the underside spots large and squarish; those of lower elevations have these spots smaller. The butterflies fly in close association with various buckwheats *(Eriogonum)* at the season when the plants are in bud, and the eggs are placed in the buds. LARVA: Variable in color and pattern, pale green or yellow with brown subdorsal markings; on flowers of buckwheat. RANGE: Scattered at a wide range of elevations, from seacoast and deserts to high-mountain habitats. A local race, the **El Segundo Blue** (**Plate 12g**), is restricted to sand dunes in Los Angeles and has been designated an endangered species by the U.S. Department of Interior.

308. Xerces or **Eyed Blue.** *Glaucopsyche xerces.* One of the natural habitats most easily affected by the expansion of civilization is the unique and delicate community of sand dunes. Such areas have always been regarded as wasteland of no value to humans so they are destroyed by planting stabilizing vegetation introduced from other parts of the world, by urban growth, sand mining or, in recent years, by off-road vehicles which extend their influence far beyond the limits of normal urban sprawl. One of the casualties was the San Francisco sand dune system, from which plants and insects were collected and described before the western expansion of the city. The Xerces Blue occurred only on the San Francisco dunes, and the last individuals were seen alive in 1943. The only species of butterfly known to have become extinct during recorded U.S. history, it has given its name to the Xerces Society, a group of conservationists hoping to prevent the disappearance of other such endangered habitats that support unique butterflies. ADULT: There were several color forms, with males typically blue-violet with a whitish or silvery sheen, females brown with a tinge of blue basally on the upperside; underneath, individuals ranged from pale gray with black-pupilled, white eyespots to dark gray with white spots; wing expanse 29–36 mm. LARVA: Color variable, pale green with a darker line down the back, distinct, yellow oblique dashes on the sides, and a pale line along each side; on *Lotus* and other plants of the pea family. An association with ants, which obtain a sugary solution from lycaenid larvae and provide feeding and protection, is known in certain other blues. This association may have been part of the delicate balance that was destroyed. RANGE: Western San Francisco peninsula; now extinct.

Plate 12h. Purplish Copper. *Lycaena helloides.* ADULT: Wing expanse 18–32 mm; upperside in male brownish with a metallic purplish sheen; with black markings and a row of reddish orange crescents on the outer margin of the HW; female (**Plate 12h**) orange without the purplish sheen and with larger, darker markings; underside in both sexes ochreous to rust-orange with faint markings. LARVA: Green with bright yellow stripes on the sides and a faint, oblique yellow dash on the side of each segment; feeds on dock *(Rumex).* RANGE: Low to middle elevations throughout the state except in the deserts. Colonies often occur in moist meadows or weedy lowland areas around cities. Adults fly from April to October.

309. Great Purple Hairstreak. *Atlides halesus.* This is California's largest and most elegant lycaenid. It is widely distributed, but usually it is not numerous and therefore all the more startling when seen. ADULT: Wing expanse 30–45 mm; male's wings are brilliant metallic turquoise blue with a black margin, a black spot on the FW, and a single thin tail on the HW; female is larger with broader, rounded, black wings that have deeper blue metallic scaling on the basal half, and the HW has two tails; underside of both sexes black with red spots at wing bases and metallic greenish spotting on the HW preceding the tails; the abdomen is bright orange underneath. LARVA: Green, velvety appearing due to covering of short, orange hairs; on mistletoe, where it is exceedingly similar in form and color to the leaves. There is no purple and nobody seems to know the origin of the common name, but tradition has resisted attempts to change it.

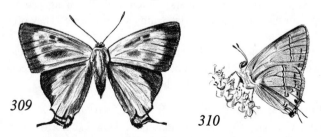

Figs. 309–310 Hairstreaks: 309, Great Purple Hairstreak, *Atlides halesus;* 310, Common Hairstreak, *Strymon melinus.*

310. Common Hairstreak. *Strymon melinus.* Also known as the **Bean Lycaenid** and the **Cotton Square Borer** owing to its adaptability to a wide range of foodplants, this species is California's most widely distributed lycaenid. ADULT: Wing expanse 20–32 mm, wings mouse gray above with a red spot on the outer, hind margin, preceding a thin, tail-like extension of the HW; underside paler, ash gray with white-margined black lines and a more extensive red and blue marginal spot. LARVA: Greenish with oblique markings of varying colors on the sides and covered with short, yellowish brown hairs; on a wide variety of plants. RANGE: Throughout lowland and foothill areas of the state, often common around cities and agricultural areas. The butterflies may be seen from February to November in southern California. A similar species, the **Avalon Hairstreak** *(S. avalona),* occurs only on Santa Catalina Island, while the Common Hairstreak occupies most of the other Channel Islands but not Catalina.

METAL MARKS
Family Riodinidae

The family Riodinidae is diverse, with many tropical members, but California has only 4 species. The butterflies are small to medium-sized with short, relatively broad wings and long antennae, measuring more than half the forewing length. The common name originates from metallic-colored markings on the wings of many species. Metal marks have reduced front legs, not used for walking in the male; those of the female, although smaller than the others, are functional. In contrast to other butterflies, metal marks perch with the wings cocked open or held out laterally, flat against the substrate. They often alight head downward on stems or undersides of leaves. The larvae are stout, with a larger head than that of Lycaenidae, and the head and legs are not hidden from view. Most species are specialized in foodplant choice.

Plate 13a. Mormon Metal Mark. *Apodemia mormo.* ADULT: Wing expanse 24–36 mm; variable in color from one geographical area to another, the wings are mostly black with orange on the inner half of the FW to mostly orange with black borders; checkered in appearance with rows of white spots; underside paler, with the FW orange, its outer margin and the HW whitish to blackish with extensive white spotting. LARVA: Purplish with rather short, gray hairs in bunches around each segment, and longer, yellowish hairs in bunches around each segment and on the sides above the legs; feeds on buckwheat *(Eriogonum)*. RANGE: In scattered colonies throughout most of the state; occupying a wide range of habitats, from beach dunes to 2500 m (above 8,000 ft) in the mountains and in desert canyons. A local race (**Plate 13a**), **Lange's Metal Mark** *(A. mormo langei)*, which is known only from the Antioch sand dunes along the San Joaquin River, is one of the first 6 insects to be designated an endangered species by the U.S. Department of Interior. The Antioch sand dunes once supported a unique community of plants and insects with desert affinities, and about 25 insects were originally discovered and described from there. Industrial development destroyed most of the dunes by the early 1950s, and subsequent stabilization of the sand by weeds and sand mining has almost eliminated the open-sand habitat. Now this butterfly is limited to 2 small colonies on lands threatened by sand removal, fires, and rototilling for fire prevention.

Family Nymphalidae

Nymphalids include a wide variety of medium-sized butterflies, all of which have the forelegs reduced and not used in walking in both sexes. The apical parts of these legs are short and clothed with long hairs, giving rise to the common name, **brush-footed butterflies.** These butterflies are broad winged, usually strong fliers, and many are known to live for several months and to travel long distances. The larvae usually are cylindrical and armed with numerous protuberances bearing spines, but they do not cause irritation to human skin. The pupae are characteristic in form for each group and often bear prominent tubercles or spines. They are suspended head downwards, hanging by hooks attached to silk buttons prepared by the larvae. A great diversity of flowering plants serve as larval food, with specialization to one or a few by most nymphalid species.

Larvae of the subfamily Danainae, represented in California by the Queen and the Monarch, are smooth with 2 or 3 pairs of fleshy filamentlike processes rather than spines. They feed on milkweed, which renders both larvae and adults distasteful to vertebrate predators. Danaines are mimicked by adults of other nymphalids which are not distasteful but gain protection from birds which have learned by experience to avoid the Danaines.

The family Nymphalidae has more members than any of the other families of true butterflies, and there are more than 50 species in California.

311. Gulf Fritillary. *Agraulis vanillae.* This species is not native in California but has long been a resident in urban areas, dependent on ornamental passion vines for larval feeding. ADULT: Wing expanse 5–8 cm; wings bright rust-orange in the male, reddish brown in the female, with black spotting; underside resplendent with elongate silver spots; FW relatively narrower than in other nymphalids. LARVA: Slate gray with deep orange stripes along the sides and black, branched spines; feeds on passion vines *(Passiflora),* especially *P. caerulea* with 5-lobed leaves. RANGE: Urban and rural areas of southern California and coastal cities of the San Francisco Bay area. In mild years the northern populations expand to inland areas where they are unable to survive winter freezing.

312; Plate 13b, 13c. Monarch. *Danaus plexippus.* Probably the world's best known butterfly, the Monarch is often used to symbolize natural beauty and butterflies in general. Its form and colors appear on everything from advertisements for bathroom tissue to floats in the New Year's Rose Parade. This species occurs throughout North America; and it is the insect best known to regularly migrate on a seasonal cycle; moving northward through spring and summer, while passing through 2 or 3 generations. The descendents migrate southward in fall to establish overwintering aggregations along the coasts and in Mexico (**Plate 13b**). On the Pacific Coast these occur in groves of trees from Sonoma County to southern California and have given rise to local protection laws in the Pacific Grove-Monterey area, where the butterflies are a tourist attraction.

Figs. 311–320 Brush-Footed Butterflies: 311, Gulf Fritillary, *Agraulis vanillae;* 312, Monarch, *Danaus plexippus;* 313, California Tortoise-Shell, *Nymphalis californica;* 314, Mourning Cloak, *N. antiopa;* 315, Painted Lady, *Vanessa cardui;* 316, West Coast Lady, *V. annabella;* 317, Red Admiral, *V. atalanta;* 318, California Sister, *Adelpha bredowi;* 319, Lorquin's Admiral, *Limenitis lorquini;* 320, Buckeye, *Junonia coenia.*

ADULT (**312**): The largest nymphalid in California (wing expanse 8–10 cm); wings rust-orange, darker in females, with black veins and black margins, marked with white dots. LARVA (**Plate 13c**): Greenish with bands of black and yellow and 2 pairs of long filaments, on the second and next to last body segments; on milkweed *(Asclepias)*. PUPA: Pale green with a thin, black belt, and raised golden spots. RANGE: Individuals may be seen almost anywhere in the state, but breeding is restricted to lower and middle-elevation areas, and overwintering to the immediate coast.

Plate 13d. Silverspots or **fritillaries.** *Speyeria*. Members of this genus are among the most familiar butterflies in mountain meadows. About 8 species occur in California; in each the wing markings and colors vary slightly from one locality to another, and the industrious lepidopterists have applied myriads of names to the differing populations. Most are similar in appearance, and 3 or 4 species may occur together. ADULTS: Wing expanse 3.8–8 cm; orange with black markings on the uppersides; HW undersides usually spangled with silver spots. LARVAE: Black or brown with lighter, lengthwise stripes which are distinct or formed of speckles; 6 rows of branching, often pale-colored spines; feed on violets. RANGE: Foothills to middle elevations of mountains throughout the state except in the desert ranges.

Plate 13e. Chalcedon or **Common Checkerspot.** *Euphydryas chalcedona*. This is one of our most abundant late-spring butterflies. ADULT: Wing expanse 3.4–6.4 cm; wings mainly black, with yellowish and red spots on the upperside, reddish orange beneath with broader, yellowish spots. In desert areas the butterflies are more reddish on the uppersides. LARVA: Black with branched spines, a row of red-orange spots along the back and variable whitish banding on the sides; feeds on Bush Monkeyflower *(Mimulus aurantiacus)* and other plants of the figwort family. Larvae often disperse to other plants when nearly full grown. PUPA: Bluish white with fine, black speckling and raised, orange spots on the abdomen. RANGE: Coastal and foothill areas throughout the state and in the desert ranges.

Plate 13f. Satyr Anglewing. *Polygonia satyrus*. This species is one of several similar-looking *Polygonia* in California. It is commonly seen in wet areas, especially shaded canyons, flying from early spring until fall. ADULT: Wing expanse 4.2–5.2 cm; wings appear ragged on the margins, orange on upperside with brown markings; underside brown with pale and darker, wavy markings, HW with a comma-shaped, silver mark in the middle. LARVA: Black with a broad stripe along the back and extensive mottling of greenish white; each segment with 7 many-branched, greenish white spines; feeds on Creek Nettle *(Urtica holosericea)* and creates a characteristic shelter by partially cutting the midrib from beneath, causing the leaf to droop downward, and the larva draws the margins together from beneath with silk. The resulting tents are always distinguishable from those of the **(317)** Red Admiral, which feeds on the same plant but creates its shelter by drawing the leaf margins upward, exposing the lowerside of the leaf. RANGE: Coastal areas and mountains to moderate elevations, wherever Creek Nettle grows; occasionally seen in cities.

313. California Tortoise-shell. *Nymphalis californica*. This species is best known for its massive outbreaks in the Mt. Shasta area, at intervals of about 5–13 years. Butterflies by the millions move outward from the mountain slopes and have been reported in newspapers to hinder motorists' visibility and to slicken highways. Following these outbreaks, individuals appear (sometimes migrating in great numbers) in distant places, as far away as British Columbia and the midwest states. In succeeding seasons these migrants sometimes give rise to temporary colonies, for example in the Coast Ranges, and as far south as San Diego, in 1952, 1960, and 1972–73. ADULT: Medium-sized (wing expanse 4.5–6 cm); ragged-margined wings, orange, broadly margined and spotted with black; underside dark brown mottled with tan and dark blue. LARVA: Black with numerous white dots each bearing a short hair and with 7 long, branched spines on each segment, those of the mid-back yellow; on *Ceanothus*. There is one generation. Adults overwinter, their larval progeny feed in spring or early summer, and the new brood of butterflies emerges from June in the Coast Ranges to early August at higher elevations, when the mass movements begin. RANGE: Probably a continuous resident only at middle elevations (1500–2500 m) (about 4500–8000 ft) in the North Coast Range, Cascades, and Sierra Nevada.

314. Mourning Cloak. *Nymphalis antiopa.* One of our most familiar butterflies. Adults are long-lived, overwinter, and occasionally fly on warm winter days so the Mourning Cloak may be seen in any month in coastal areas. ADULT: Wing expanse 6–9 cm; wings rich brownish black, tinged with deep, lustrous purplish when fresh, with a yellow border preceded by a row of pale blue spots; underside mottled blackish brown with a dull, yellowish border. LARVA: Feeding in a group when young, later dispersing and more or less solitary; black with 7 rows of long, thin, spines, white speckling, and a row of dull orange spots along the back; lives on willow, elm, alder, and poplar. RANGE: Throughout the state except at the highest elevations; in the deserts only in willow-lined mountain canyons.

315. Painted Lady. *Vanessa cardui.* One of the best known, least variable, and most cosmopolitan butterflies in the world, the Painted Lady periodically migrates in large numbers northward throughout California. Northern colonies, however, usually die out during winter. ADULT: Wing expanse 4.2–6.5 cm; similar in appearance to the (**316**) West Coast Lady but larger, FW with rounded margins and without the blue-pupilled eyespots of the HW. LARVA: Lilac colored with black spots and yellow, black, and white lines separating the segments, with the branched spines short and dark; on thistle *(Cirsium),* mallow *(Malva),* and other low-growing plants. In years when great population outbreaks occur in the deserts, a wide variety of plants is used by successive northward-moving generations. RANGE: Probably a permanent resident only in the deserts and coastal areas of southern California.

316. West Coast Lady. *Vanessa annabella.* Long known as *Vanessa carye,* this species is very similar to the following, except in color; occasional natural hybrids of the 2 are known.

The West Coast Lady is more often seen in numbers, especially in weedy areas. ADULT: Wing expanse 3.8–5.5 cm; wings orange with black markings, HW with a row of 4 black-bordered, pale blue spots; FW underneath rose-pink with blue-gray and whitish markings; HW dull colored, mottled in bluish and grays. LARVA: Variable in color, usually blackish with orange blotches, covered with fine hairs and with 7 pale, branched spines on each segment; on mallow *(Malva)* and related plants and on nettle *(Urtica)*. The caterpillar forms a cup-shaped "nest" by pulling the margins of the leaf upwards with silk. RANGE: Throughout most of the state at low to middle elevations, scarce in the deserts.

317. Red Admiral. *Vanessa atalanta.* This species is widely distributed but not usually observed in large numbers. It is most often seen in cities during fall months. ADULT: Wing expanse 4.4–6 cm; wings black above with a transverse, orange band followed by white spots in apical part of FW and a broad, orange border on HW; underneath the transverse band is pink on the FW, and the wings are mottled with brown and blue. LARVA: Black studded with whitish granulations and 7 light-colored, branched spines on each segment; feeds on nettle *(Urtica)* and related plants. RANGE: Throughout most of the state at a wide range of elevations, although scarce in the desert ranges. The butterflies may be seen throughout the year in coastal areas.

318. California Sister. *Adelpha bredowi.* This is one of California's most easily recognized butterflies, with its broad wings and characteristic sailing flight as it glides among oak trees. ADULT: Wing expanse 5.5–8 cm; wings black with a transverse band of white spots and a broad, orange spot before the tip of the FW; underneath these markings are less distinct and the wings have blue and reddish bands and spots. LARVA: Dark green above, paler below, with 6 pairs of green, fleshy tubercles; on oaks. RANGE: Foothills and middle elevations of mountains throughout the state.

319. Lorquin's Admiral. *Limenitis lorquini.* This species re-sembles the California Sister, but does not have its gliding flight. The Lorquin's Admiral occurs along willow-lined creeks, where males establish perch stations and dash out at passing insects or thrown objects. ADULT: Similar to the preceding species but FW with a narrower orange patch extending to the tip; wing expanse 5–7 cm. LARVA: A bizarre creature resembling a bird dropping, olive brown with yellowish markings on the thorax and a whitish, mid-abdominal patch on the back and 2 fleshy protuberances extending forward from the second segment; lives on willow, poplar, and occasionally trees of the rose family, including apple and wild cherry. RANGE: Coastal areas, inland valleys, and mountains to moderately high elevations.

320; Plate 13g. Buckeye. *Junonia coenia.* The Buckeye is familiar to most naturalists from its distinctive appearance. It perches with its wings spread and flies throughout the year in warmer areas, being most abundant in fall. ADULT (**320**): Wing expanse 3.5–5.2 cm; wings brown, marked on the uppersides by a large and small eye spot on each wing and by basal orange bars and a broad white blotch on the FW; underside of FW similar to upperside, mottled in dull brown, red-brown, or olive on HW. LARVA (**Plate 13g**): Dark with longitudinal stripes of pale yellow and short, branching spines; feeds on plantain *(Plantago)* and related plants and on members of the figwort family. RANGE: Throughout the state at a wide range of elevations, scarce in the deserts.

Family Satyridae

Satyrids are medium-sized, brown or whitish butterflies with round-margined wings that have some of the veins greatly swollen at the base. Most species have some form of eyelike spots, at least on the undersides. The butterflies perch with the wings closed, and the eyespots are believed to provide protection from predators by deflecting the point of attack from vital body parts. The flight is weak and of a characteristic bouncing gait, unlike that of other butterflies. Satyrids usually seek sheltered spots on the ground where they perch motionless as a means of escape. The larvae are spindle-shaped, tapering at each end. The head is often horned, and the tail end has a pair of short, backwardly directed processes. The body otherwise lacks spines or protuberances. Most species feed on grasses.

This family is primarily a temperate zone or boreal group, often restricted to mountainous regions, with 7 species occurring in California.

321. California Ringlet. *Coenonympha california*. One of the state's most ubiquitous and abundant butterflies, this species is sometimes overlooked owing to its low flight and mothlike appearance. ADULT: Wing expanse 24–38 mm; wings whitish above, mottled with gray or tan beneath; HW

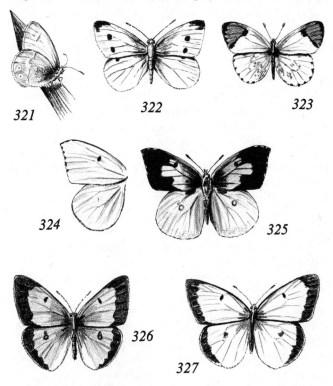

Figs. 321–327 Butterflies: 321, California Ringlet, *Coenonympha california;* 322, Cabbage Butterfly, *Pieris rapae;* 323, Sara Orange-Tip, *Anthocharis sara;* 324, California Dogface, *Colias eurydice* (female); 325, *C. eurydice* (male); 326, Alfalfa Butterfly, *C. eurytheme;* 327, Clouded Sulfur, *C. philodice.*

with small, circular ring spots. The colors and number of spots vary seasonally and geographically, with darker forms in early spring and ochreous forms in summer and east of the Sierra Nevada. LARVA: Olive green or brownish, marked with light and dark longitudinal lines; on grasses, apparently including introduced species. RANGE: Foothill and oak woodland areas throughout California, outside the deserts.

Family Pieridae

Included in the family Pieridae are some of our most familiar butterflies, **whites** and **sulfurs.** They are mostly medium-sized, yellow or white with black markings, and are strong fliers, occurring for the most part in open areas. The larvae are rather elongate with the segments divided by creases into ringlike sections; the body is covered with numerous short hairs but has no spines or protuberances. Most pierids feed on plants of either the mustard or pea families. Pierids are diversified in all regions of the world. About 25 species are resident in California, occurring in a wide range of habitats from the deserts and coasts to timberline.

322. Cabbage Butterfly or **Imported Cabbageworm.** *Pieris rapae*. Probably this is the most injurious butterfly to man's economy. Originally widely distributed in the Old World, it was introduced into North America more than a century ago and now occurs throughout the United States in association with gardens, truck crops, and weed fields. ADULT: Wing expanse 38–52 mm; white, FW with black tips; male has a single black spot on each wing, female has 2 on the FW, 1 on the HW; underneath yellowish, FW with upperside spots repeated. LARVA: Pale green, finely dotted with black, with a yellowish stripe down the back and a row of yellowish spots on each side; feeds on all kinds of cultivated and weedy plants of the mustard family. RANGE: Throughout low- to middle-elevation parts of California, although scarce in the deserts. The adults fly throughout the year in coastal areas.

323. Sara Orange-tip. *Anthocharis sara*. This is one of our commonest and most distinctive spring butterflies. ADULT: Wing expanse 28–46 mm; male white with the apical one-third of the FW reddish orange, bordered by black; female either white or yellow, with a paler, less well-defined, orange tip; beneath, wings heavily mottled with dark green in the early spring generation, less strongly so in the late spring brood which consists of larger but less numerous individuals. LARVA: Green with a wide, yellowish stripe along each side above the legs, bordered below by a darker green area; back mottled with black dots with an indistinct darker line; feeds on various plants of the mustard family. RANGE: Canyons of foothills to middle-elevation mountainous areas throughout the state. A small, dark race occurs in the mountains of the Mojave Desert.

324, 325. California Dogface. *Colias eurydice.* This is California's official state butterfly, so designated by the State Legislature in 1972. ADULT: Wing expanse 48–65 mm; male (**325**) has a yellow silhouette of a dog's head, defined by black over the outer half of the FW; fresh specimens have a beautiful violet iridescence; HW yellow with or without a black border; female (**324**) usually is entirely yellow with a black spot at the position of the dog's eye, but occasional specimens have clouded blackish markings that poorly define the dog's head; undersides variable, from ochreous to pinkish. LARVA: Dull green with a white lateral line, edged below with orange; feeds on False Indigo *(Amorpha californica)*. RANGE: Foothills of the Sierra Nevada and Coast Ranges from Sonoma to San Diego counties; more abundant in the south. A similar species, the **Southern Dogface** *(C. cesonia),* occurs in the deserts. It lacks the violet iridescence, and females normally are marked with the dog's head pattern.

326, 327. Alfalfa Sulfur or **Alfalfa Caterpillar.** *Colias eurytheme.* This butterfly swarms around alfalfa fields in agricultural areas and occurs widely in natural and weedy situations. ADULT: Wing expanse 32–55 mm; male is yellow-orange with a distinct black border on the upperside of the wings; females have 2 forms, white and yellow-orange, and have clouded blackish markings. The amount and intensity of orange are seasonally variable. LARVA: Dark green with a white line on each side above the legs that borders a narrow, reddish line; on alfalfa, clovers, and other plants of the pea family. RANGE: Almost everywhere in the state, even at high elevations. The closely similar (**327**) **Common** or **Clouded Sulfur** *(C. philodice)* lacks an orange tinge to the yellow ground color. It is a more widespread species through the United States but occurs in California only in Great Basin areas of the Modoc Plateau and east of the Sierra Nevada and in the Imperial Valley.

Family Papilionidae

The Papilionidae includes the **swallowtails** and **parnassians,** some of the most magnificent of all insects. Most species are large with showy colors; although variable in form, the hindwings of most have conspicuous, tail-like extensions at the end of one or more veins. Larvae of swallowtails are smooth or provided with fleshy tubercles, while those of parnassians have clumps of hairs arising from raised spots. Papilionid larvae have characteristic organs called osmeteria, that consist of a forked sac which is protruded through a slit on the back of the first body segment. When disturbed, the larva everts the osmeterium and emits a strong odor, varying according to species, and often extremely disagreeable to mammals. The horn-shaped organ usually is bright orange or pink, and contrasts sharply with the cryptic color of the larva. The pupa or chrysalis of swallowtails is also cryptic in form and color. It is formed head-upwards and held by a girdle of silk at the middle and a button of silk at the tail. Parnassians pupate in a frail silken cocoon among leaves. Usually a few related plants are fed upon by larvae of any given species. In many cases the plants render the larvae and adults distasteful to vertebrate predators, so papilionids often serve as the models for mimicry by other insects, especially in the tropics.

Parnassians are large, white butterflies having tailless, rounded wings with black and orange spotting. Two species occur in California and are sometimes common in Sierran meadows. The larvae are believed to feed on stonecrops *(Sedum)* and bleeding heart *(Dicentra)*.

Most papilionids are tropical, with only a small representation reaching North America north of Mexico. Only about a dozen species occur in California.

Plate 13h; cover photo (lower left). Anise Swallowtail. *Papilio zelicaon.* This is our commonest swallowtail around cities. ADULT (**cover photo**): wing expanse 6.5– 9 cm; yellow with black markings; black band on HW with yellow marginal crescents, preceded by bluish crescents, ending in an orange-margined eyespot at the inner corner; underside with considerable orange scaling in the yellow areas. LARVA (**Plate 13h**): First 3 stages black with a white blotch on the back, resembling bird droppings; last 2 stages green, striped with black on each segment, with orange spots in the black bands, resembling plant colors, especially when feeding on flower heads; feed on Anise or Sweet Fennel *(Foeniculum)* and other members of the parsley family and on citrus. RANGE: Throughout most of the state at a wide range of elevations, but most abundant at low elevations near the coast. In southern California the butterflies may be seen throughout the year.

328. Western Tiger Swallowtail. *Papilio rutulus.* This species is also a familiar sight around cities, especially along willow-lined creekbeds. ADULT: Wing expanse 7–10.5 cm; yellow with black markings; the black HW margin contains yellow crescents, faint bluish crescents towards the tail end, and a crescent of orange at the hind corner. LARVA: Deep green with a yellow line, edged with black back of the fourth body segment; has 2 yellow, black-bordered spots pupilled with blue; these resemble eyes when the larva is at rest with the head and first 2 segments tucked under; on willow, poplar, and

Figs. 328–329 Swallowtails: 328, Western Tiger Swallowtail, *Papilio rutulus;* 329, Pipevine Swallowtail, *Battus philenor* (underside).

sycamore. RANGE: At low to middle elevations throughout the state except in the deserts. A similar species, the **Pale Swallowtail** *(P. eurymedon),* has whitish rather than bright yellow background color. It is less common but often occurs in the same habitats as the Tiger Swallowtail.

329. Pipevine Swallowtail. *Battus philenor.* ADULT: Wing expanse 6–9.5 cm; blackish with metallic greenish blue and a row of white spots on the HW; beneath, the wings are magnificently colored in metallic blue-green with a row of bright orange and white spots on the HW. LARVA: Black with bright reddish orange tubercles; on Dutchman's Pipevine *(Aristolochia californica).* The larvae and adults are protected from vertebrate predators by distastefulness originating from the foodplant. RANGE: Common in foothills of the Coast Ranges and Sierra Nevada in the northern half of the state, where the butterflies are smaller and densely hairy. The larger, typical form is taken occasionally in southern California, but no pipevine is known there, and these may be wandering individuals from populations in Arizona.

BEETLES
Order Coleoptera

Beetles make up the largest order in the animal kingdom, with more than twice the number of described species of the next largest order, Hymenoptera. Coleoptera are unsurpassed in morphological modifications and behavioral adaptations. Beetles occur in virtually every conceivable habitat. They especially inhabit the ground and either live in the soil or on it, using decaying animal or vegetable substances. Dung, carrion, refuse, humus, rotting wood, and fungi all support large assemblages of Coleoptera. A particularly rich habitat is beneath bark of dead, standing or fallen trees, where some species of about half the families can be found. Moreover, many species eat living plants, and their larvae are found on or in flowers, fruits, leaves, stems, or roots. A few are parasitic on other insects, and many are predators. Twelve families are true aquatic insects and other families have aquatic or semiaquatic representatives. There are at least 7,000 species in California, and many await discovery, particularly in families where adults are small.

Included in the order are some of the largest and some of the smallest of all insects. Usually they have 2 pairs of wings, with the forewings modified into hard or leathery covers (elytra), which almost always meet to form a straight line down the back. This is the most easily seen characteristic of Coleoptera; however, many species have the hindwings reduced or missing and are flightless. When present, the hindwings are membranous and folded beneath the elytra when not used for flying. The mouthparts are adapted for biting and chewing, with well-developed mandibles. The metamorphosis is complete and larvae show a great diversity of form and habits: grublike borers, rapidly running or swimming, silverfish-shaped predators, caterpillarlike plant feeders, and so on. Normally pupation takes place in chambers in the soil or wood, without a cocoon.

References

Arnett, R. H. 1963. *The Beetles of the United States, a Manual for Identification*. Washington, D.C.: Catholic University Press.

Bright, D. E., and R. W. Stark. 1973. The bark and ambrosia beetles of California (Coleoptera: Scolytidae). *Calif. Insect Survey, Bull.* 18; 169 pp.

Hatch, M. H. 1953–1971. *The Beetles of the Pacific Northwest*. Seattle: University of Washington Press; 5 vol.

Jaques, H. E. 1951. *How to Know the Beetles*. Dubuque, Iowa: W.C. Brown Co.

Leech, H. B., and H. P. Chandler. 1956. Coleoptera. In *Aquatic Insects of California*, ed. Usinger, pp. 293–371. Berkeley and Los Angeles: University of California Press.

Linsley, E. G., and J. A. Chemsak. 1961–1976. The Cerambycidae of North America, Vols. 1–6 (continuing). *Univ. California Publ. Entom.*, 18–22, 69, 80.

TIGER BEETLES
Family Cicindelidae

Tiger beetles are among the most voracious of all insects. Adults stalk their prey on the ground, and the larvae wait in burrows to snatch passing victims. The beetles are characterized by prominent eyes with the antennae arising between them, and by long legs armed with spines. Members of the genus *Cicindela* are metallic-colored, rapid fliers and are able to run down quick insects on the ground; those of *Omus* are slow-moving, flightless, nocturnal, and black. In both genera, the larvae are similarly modified; the fifth abdominal segment is swollen and bears a pair of hooks on the back which, along with the bent form of the larva, serves to hold it in its burrow when prey is grasped. The head and first segment of the thorax are broadened into a roughened shield which is held at ground level like a trap-door when the larva is waiting for prey. There are 30–40 species of cicindelids in California.

330. California Black Tiger Beetle. *Omus californicus.* ADULT: BL 13–19 mm; flightless, with the elytra fused. Black without markings; thorax roughened. LARVA: Similar to that of *Cicindela* but larger when full grown; in burrows in hard-packed canyon banks and road beds in forested areas. RANGE: Coastal areas, Coast Ranges and Sierra Nevada to middle elevations.

331. Oregon Tiger Beetle. *Cicindela oregona.* This and related species inhabit sandy areas around water. They are exceedingly quick, wary, and hard to follow when they fly, defying the inexperienced collector. ADULT: BL 10–14 mm; upperside dull olive brown with whitish, variable markings; underside shining metallic green. LARVA: Whitish with the swollen fifth segment brown and the prothorax sand colored; lives in burrows in sand along sea beaches, creeks, seepages, and lake shores. RANGE: Coastal and mountainous parts of the state and along Central Valley streams.

PREDACEOUS GROUND BEETLES
Family Carabidae

The Carabidae is one of the most diverse families of Coleoptera in numbers of species and in size. The beetles usually are dark-colored with shiny wing covers and long legs. Carabids are similar to tiger beetles but have the antenna arising from the side of the head between the eye and the base of the mandible. Adults of many species have the wing covers fused and hindwings atrophied. They are active predators by

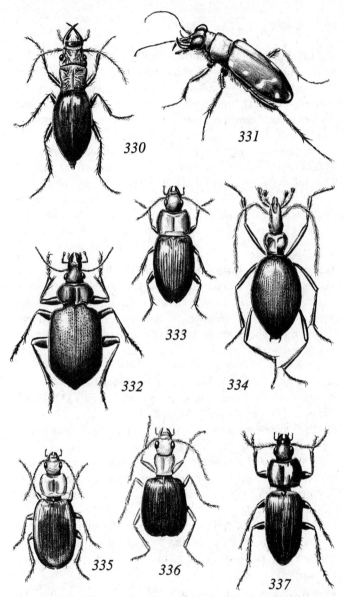

Figs. 330–337 Beetles: 330, California Black Tiger Beetle, *Omus californicus;* 331, Oregon Tiger Beetle, *Cicindela oregona;* 332, Black Calosoma, *Calosoma semilaeve;* 333, Rufous Carabid, *Calathus ruficollis;* 334, Snaileater, *Scaphinotus* species; 335, Tule Beetle, *Agonum marginicolle;* 336, Bombardier Beetle, *Brachinus tschernikhi;* 337, Common Carabid, *Pterostichus californicus.*

night and generally hide during the day under stones or other objects, especially in moist areas. Owing to this habit and their diversity, beginning collectors always have a variety of species. Carabids are commonly found in gardens, around compost heaps and trash piles, etc., wherever other ground-dwelling, insect prey may be found.

The larvae are active predators with an elongate and somewhat flattened form. They have hardened plates on the back, at least on the thorax, and have a prominent head and long, curved mandibles. They live under stones, in moss, rotting wood, under bark, and other similar habitats. There is a tremendous diversity of carabids in temperate zones, and probably there are over 800 described species in California.

332. Calosomas. *Calosoma.* Members of this genus are large beetles having the thorax nearly as wide as the abdomen, much wider than the head. They are black, green, or bronze colored, often with a metallic iridescence. Some species are expert tree climbers and voracious feeders on caterpillars and other insect defoliators. Adults are often seen at lights, feeding on remains of insects that have been crushed underfoot. When disturbed, these beetles give off a distinctive, fetid odor from abdominal glands. One of the commonest in California is the (**332**) **Black Calosoma** *(C. semilaeve).* ADULT: Large (BL 20–27 mm), black, and almost smooth. LARVA: Shining black above with white markings on the sides and venter. Feeds on cutworms and other ground-inhabiting larvae. RANGE: Low elevations in coastal southern California, Imperial and Central valleys. California's most spectacular carabid is the **Green Hunter** *(C. scrutator),* a large (BL 28–34 mm), steel blue species with burnished head and thorax and metallic green elytra margined in gold. The beetles live on the ground but are said to commonly climb willows to catch buckmoth caterpillars in the Central Valley.

333. Rufous Carabid. *Calathus ruficollis.* ADULT: BL 7–10 mm; shiny with the thorax short but nearly as wide as the abdomen; reddish brown with thorax brighter reddish than the finely lined wing covers. Common in many habitats in the San Francisco Bay area. RANGE: Coastal areas and Coast Ranges.

334. Snaileaters. *Scaphinotus.* These beetles have the head and mouthparts adapted for entering the opening of a snail shell and scooping out its contents. They live in cool, damp woods and gardens where they hunt at night on logs, tree trunks, and on the ground. ADULTS: Large (BL 14–22 mm); black or reddish brown, some species with purplish iridescence; characterized by a small thorax and large abdomen. LARVAE: Dark brown or black, heavily armored, with sharp jaws and long antennae. RANGE: Coastal areas, Coast Ranges, and Sierra Nevada.

335. Tule beetles. *Agonum.* These beetles are similar to the (**333**) Rufous Carabid and are often the commonest carabids under rocks and debris. ADULTS: Small (BL 8–12 mm); shiny with a rather narrow, rounded thorax. (**335**) *A. marginicolle* is pale brown with a black blotch in the middle of the thorax and all but the margins of the elytra black. Commonly breeds in marshy places, often appearing in swarms after fall rains. The beetles enter houses and are annoying because of their offensive odor. RANGE: Low-elevation areas along the coast, especially around rivers, very common in the Central Valley. *A. funebre* is similar but is entirely shining black. It is widespread in coastal southern and central California and in the Sierra Nevada.

336. Bombardier Beetle. *Brachinus tschernikhi.* This species lives at the rocky margins of lakes and streams. When disturbed, *Brachinus* produces a chemical from glands at the tail end, which mixes in an internal "combustion chamber" and is expelled as a visible puff of steam that is irritating to would-be predators. The explosion is accompanied by a popping or hissing noise. ADULT: Small (BL 7–12 mm); reddish orange with dark, metallic blue wing covers etched with parallel lines. LARVA: Blackish with reduced legs, an adaptation for parasitic life in pupal cells of stream-inhabiting beetles. RANGE: Low-elevation riverbeds throughout the state. The similar beetles of the genus *Chlaenius* often occur with *Brachinus* in the same habitat. *Chlaenius* are larger (BL 12–16 mm); with dark metallic blue body and red or orange legs.

They produce a defensive secretion but not as a vaporous, noisy explosion.

337. Common carabids. *Holciophorus, Pterostichus.* ADULTS: BL 10–28 mm; moderately elongate, thorax nearly as broad as the abdomen; with fine lines etched along the length of wing covers. *H. ater* is large (BL 24–28 mm); occurs in the north coast area and Sierra Nevada. *P. vicinus* and **(337)** *P. californicus* are smaller, black, and are very common in the San Francisco Bay region and in the Sierra Nevada. *P. brunnea* is dark brown and often occurs in drier habitats, on the east side of the Coast Range and elsewhere.

PREDACEOUS WATER BEETLES
Family Dytiscidae

Members of the family Dytiscidae live in ponds and streams and are carnivorous in both larval and adult stages. The beetles are streamlined, oval in outline, and convex both above and below. They are distinguished from the scavenger water beetles, which they most resemble, by having threadlike antennae, longer than the head. Dytiscids swim by simultaneous strokes of the hindlegs, which are flattened and fringed with long hairs. They breathe by periodically coming to the surface to collect a bubble of air which is taken at the tail end and held beneath the wing covers. The adults of most species can fly from pond to pond, and they are sometimes attracted to lights.

The larvae are slender, spindle-shaped, often strongly tapered towards the head and tail. They swim with the aid of the legs which are fringed with hairs. Breathing occurs through a pair of pores at the tail end; the last abdominal segments are fringed with hairs, enabling individuals to hang downwards from the surface film. Some species have long, lateral filaments on the abdominal segments, which do not function as gills.

This family is well represented in temperate regions, with about 125 species known in California. Dytiscids are widespread, numerous in species, often abundant, and active during the day. With the possible exception of the whirligig beetles, they are more often noticed than other aquatic Coleoptera, and most people who refer to "water beetles" have dytiscids in mind.

338. Yellow-spotted water beetles. *Thermonectus.* Two pond-dwelling species include the most colorful of California's water beetles. ADULTS: Medium-sized (BL 10–15 mm); rather broad, oval in outline. (**338**) *T. marmoratus* is black with 10 or 11 spots on each wing cover. The spots appear bright orange when the beetles are in the water, but the color fades to dull yellow after specimens die. RANGE: Riverside and San Diego counties. *T. basillaris* is blackish with irregular, longitudinal, yellowish marks on the sides and with yellowish legs. RANGE: Ponds in drier portions of the state at lower elevations from Shasta County south to San Diego.

339. Giant Green Water Beetle. *Dytiscus marginicollis.* This is the largest dytiscid in California, and the adults often are attracted to lights, where the beetles clumsily thrash about, attracting attention. ADULT: Large (BL 27–30 mm); with smooth, dark green wing covers and the margins of the thorax and wings dull yellow. LARVA ("water tiger"): Large (BL 65–70 mm); yellowish with greenish brown dorsal plates; tubular with the abdomen gradually tapered; head broad, depressed, with long, slender mandibles. Adults and larvae eat aquatic insects, tadpoles, and small fish and occasionally are destructive in hatcheries. RANGE: Coastal areas and Central Valley. The **large elliptic water beetles,** *Cybister ellipticus* and *C. explanatus,* are similar but lack the yellow margin around the thorax and have the wing covers broadest beyond the middle. The latter species occurs in the Central Valley, Owens Valley, and southern California, while *C. ellipticus* is limited to San Diego and Imperial counties.

340. River beetles. *Agabus.* Members of this genus often are the most prevalent water beetles in fast-flowing streams and rivers. There are about 25 species in California. ADULTS: Small to medium-sized (BL 7–12 mm); mostly black or with pale to dark brown wing covers, usually without longitudinal

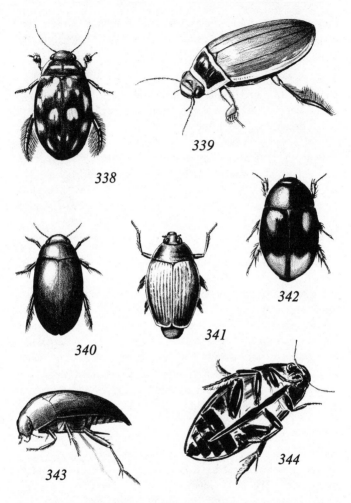

Figs. 338–344 Water Beetles: 338, Yellow-spotted Water Beetle, *Thermonectus marmoratus;* 339, Giant Green Water Beetle, *Dytiscus marginicollis;* 340, River Beetle, *Agabus* species; 341, Whirligig Beetle, *Gyrinus* species; 342, Spotted Dung Beetle, *Sphaeridium scarabaeoides;* 343, Scavenger Water Beetle, *Tropisternus* species; 344, Giant Black Water Beetle, *Hydrophilus triangularis* (underside).

lines; margin of each eye with an indentation above the antenna. RANGE: Throughout the state wherever permanently running streams occur.

WHIRLIGIG BEETLES
Family Gyrinidae

Whirligig beetles are perhaps the most unmistakably recognizable family of all Coleoptera: shiny, streamlined beetles that swim on the surface of ponds and still parts of streams. They are most often seen in small aggregations whirling in amazing feats of tireless acrobatic skating, constantly darting in graceful curves around one another.

All the species are similar in appearance, being elongate-oval in outline, somewhat flattened, and steely black. They are otherwise easily distinguished from other water beetles by the eyes, which are divided into upper and lower portions, enabling them to see both above and below the water surface. The hindlegs, and middle legs to a lesser degree, are flattened, adapted for swimming. The larva is elongate with deeply constricted segments. Each of the first 8 abdominal segments bears a pair of lateral, plumose gills, and 2 pairs are borne on the ninth segment. Both larvae and adults are predaceous on other small aquatic insects, worms, or small vertebrates. About a dozen species are known in California.

341. Common whirligig beetles. *Gyrinus.* Most of the state's gyrinids are similar-appearing members of this genus. ADULTS: Small (BL 5–6 mm); shining black. RANGE: Throughout the state at a wide range of elevations, from desert canyons to above 2,000 m (6500 ft) in the Sierra Nevada.

SCAVENGER WATER BEETLES
Family Hydrophilidae

The family Hydrophilidae includes beetles of diverse sizes and shapes, mostly adapted for life in stagnant ponds, shallows of lakes, or quiet places in streams containing an abundance of vegetation. Some species live in moist, decaying leaves, soil rich in humus, or fresh mammal dung. Hydrophilids are distinguished from other water beetles, especially the dytiscids which they most resemble, by having short, clubbed antennae held in small grooves next to the eyes. The maxillary palpi usually are elongate and resemble antennae to the casual glance. The aquatic species are generally not so well adapted for swimming as are the predaceous water beetles, being less streamlined, with the back more convex, and the underside concave, often bearing a conspicuous, posteriorly directed, median spine or keel. They swim by alternately stroking the hindlegs and breathe by breaking the surface film with the hairy antennal club. Air taken in along pubescent tracts on either side of the thorax is held under the wing covers and in hairy tracts along the underside.

The larvae also are of diverse form and structure: some of the aquatics have long, gill-like structures on the first 7 abdominal segments, but in most the breathing pores are functional only on the last segment, occurring there on protrusions or in pits; in terrestrial species the larvae are grublike with vestigial legs. The adults are scavengers on decaying organic material, although a few are predaceous. The larvae of most are predaceous on other invertebrates.

Most hydrophilids have dispersal flights in spring or fall. At these times the beetles sometimes are attracted to lights in great numbers. The family is cosmopolitan, and about 70 species representing 17 genera are recorded in California.

342. Spotted Dung Beetle. *Sphaeridium scarabaeoides.* ADULT: Small (BL 5–7 mm); circular in outline with convex back; blackish with yellowish orange tips and a reddish spot on each wing cover. LARVA: Somewhat C-shaped with the underside on the outer curve, grublike with short legs, not tapered posteriorly; last segment enlarged, with a pair of lateral processes on each side; feeding on fly larvae in mammal dung. RANGE: Throughout the state wherever grazing occurs. The beetles occur in fresh dung, often when it is of almost liquid consistency.

343. Common scavenger water beetles. *Tropisternus.* Several species of this genus are among the larger aquatic hydrophilids in California. ADULTS: Medium-sized (BL 8–12 mm); upperside strongly arched, shining black or greenish black, with smooth thorax and wing covers; a keel projecting beyond the legs on underside. LARVAE: Rather stout, spindle-shaped but not strongly tapering towards the tail; with short processes on the sides and back of each abdominal segment. RANGE: Throughout the state at a wide range of elevations in ponds including coastal saline lagoons in southern California.

344. Giant Black Water Beetle. *Hydrophilus triangularis.* The largest and best known member of the family, this beetle often comes to lights, attracting attention as it thrashes about noisily but harmlessly. ADULT: BL 33–38 mm; shining black with a prominent keel on the underside. LARVA: Elongate, somewhat spindle-shaped, strongly tapering towards the tail end; without lateral, gill-like processes; head strongly pigmented; body without armor plates. Both adults and larvae are predaceous, devouring all procurable dead and living animals and are said to be troublesome in fish hatcheries at times. RANGE: Widespread; east of the Sierra Nevada, central coastal areas, and particularly common in the Central Valley.

ROVE BEETLES
Family Staphylinidae

Rove beetles are easily recognized by their short wing covers. In most species the abdomen is almost completely exposed and often it is curled upwards in a characteristic fashion when the beetles are running about. Usually the hindwings are large. Although completely folded beneath the short elytra, they can be unfolded quickly for instant flight. Staphylinid larvae are silverfish-shaped, resembling those of predaceous ground beetles (Carabidae), often are armored with sclerotized plates on the back, and have a tubular tail-end segment bearing 2 segmented tails (cerci). Many are predaceous and some are parasitic inside pupae of other insects, but the habits of most species await investigation.

Rove beetles occur in diverse habitats, abounding wherever there is decaying organic material, including dung and animal carcasses. Some live in fungi, others along streamsides, or even in intertidal zones on the ocean shore, and a large number are ant associates, especially in tropical regions. There is a tremendous diversity of species, with nearly 1,000 described in California and probably many others awaiting discovery. The majority are small and inconspicuous; the 3 illustrated are our largest species.

345. Hairy Rove Beetle. *Staphylinus maxillosus*. This species occurs around dead animals where the adults and larvae feed on maggots of carrion flies. ADULT: Large (BL 12–22 mm); black, densely clothed with hairs that form distinct yellow or white markings on the wing covers and basal abdominal segments. RANGE: Widely distributed in California at lower to middle elevations.

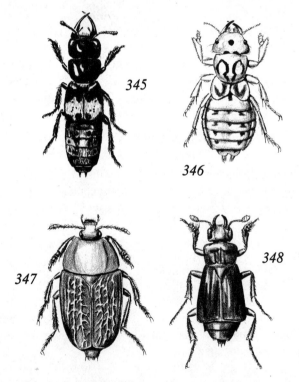

Figs. 345–348 Beetles: 345, Hairy Rove Beetle, *Staphylinus maxillosus;* 346, Pictured Rove Beetle, *Thinopinus pictus;* 347, Garden Silphid, *Silpha ramosa;* 348, Black Burying Beetle, *Nicrophorus nigritus*.

Plate 14a. Devil's Coach Horse. *Ocypus olens.* The largest staphylinid in our fauna, this species was introduced from Europe and was first recorded in California in 1931. ADULT: BL 17–32 mm; black, without conspicuous hairiness. A foul-smelling liquid is emitted from 2 cream-colored glands at the tip of the abdomen when the insect is alarmed. LARVA: Predaceous on slugs and snails. RANGE: Widely established in urban areas of southern California and the San Francisco Bay area.

346. Pictured Rove Beetle. *Thinopinus pictus.* This remarkable beetle lives on beaches where it may be found in great numbers under seaweed and on open sand at night. ADULT: Large (BL 12–22 mm); robust, broad, pale ochreous brown with blackish markings on the back. Unlike most staphylinids, it is wingless. LARVA: Similar to the adult, narrower without wing covers, and with the second and third thoracic segments shining, dark red-brown. Both adults and larvae are predaceous on small organisms, especially sand fleas (amphipods). RANGE: Beaches along the length of California.

CARRION BEETLES
Family Silphidae

Silphids are flattened to robust with the wing covers slightly shorter than the abdomen. They have conspicuous antennae that are enlarged gradually or terminate abruptly in a flattened club. Nearly all feed on carrion, although some are vegetarians. The larvae of most silphids are somewhat flattened with a shieldlike back which covers the head and legs, but in burying beetles they are grublike. The family is small, primarily of Temperate Zone distribution, with about 15 species in California.

347. Garden Silphid. *Silpha ramosa.* ADULT: BL 12–18 mm; flattened; black with 3 irregular, longitudinal ridges on each wing cover. LARVA: Resembling a sowbug, black with the dorsal plates extended laterally to form a shield that hides the head and legs from view. Adults and larvae are general feeders, consuming living or dead, soft-bodied insects such as maggots, which feed on decaying organic matter in the soil;

Silpha sometimes feed on leaves of plants at night. RANGE: Widely distributed in California at lower elevations. A similar species, the **Satin Silphid** *(S. lapponica),* is smaller (BL 8–15 mm); with an olive-colored, velvety sheen of scales on the head and thorax. It occurs at a wide range of elevations and is more common in native habitats.

348. Burying beetles. *Nicrophorus.* The adults dehair and macerate carcasses of small animals, such as rodents and snakes, and bury them in chambers. This helps prevent fly larvae from eating the carrion and slows the drying of the corpse until the silphid larvae mature. The eggs are laid in a depression on top of the carcass, and the beetles remain with the larval brood until it matures in about a week. ADULTS: Moderately large (BL 12–28 mm); shining black with yellow or orange markings or antennae. These beetles commonly carry colonies of large mites under the wing covers. LARVAE: Grublike, somewhat flattened, broadly spindle-shaped, broadest at the anterior abdominal segments; with a row of blunt spines across the back of each segment. RANGE: Coastal and mountainous parts of California at a wide range of elevations. The **Red and Black Burying Beetle** *(N. marginatus)* is one of our common species. It has the antennal club, stripes on the sides, and 2 wide bands across the wing covers, orange or red. The **(348) Black Burying Beetle** *(N. nigritus)* is entirely black.

STAG BEETLES
Family Lucanidae

Stag beetles derive their common name from the great development of the mandibles or "horns" on the heads of males. Males are usually larger than the females and show a greater variation in size. These variations are coupled with differences in development of the head; larger beetles have disproportionately larger mandibles or horns. The antennae terminate in broad, flattened segments that cannot be closed together to form a compact club as in the Scarabaeidae. Lucanid larvae are C-shaped grubs that make rasping noises by rubbing hard ridges on the third pair of legs against a rugose area at the base of the second pair. They inhabit soft, rotting wood of fallen trees and stumps, and 4–6 years may be required to reach maturity. Often the adults are found with the larvae, and the beetles are rarely seen away from wood. This is a small family in temperate regions with less than 10 species known in California.

349. Rugose Stag Beetle. *Sinodendron rugosum.* ADULT: More robust than the Oregon Stag Beetle, cylindrical; BL 10–18 mm; head much narrower than thorax; black; rugose; the head in the male with a short horn that curves backward. LARVA: Head brown, body dirty whitish with the dark gut contents showing through, the last 3 segments enlarged; in rotten wood of oak, California Bay, and other trees. RANGE: Coast Ranges and Sierra Nevada to middle elevations.

350. Oregon Stag Beetle. *Platycerus oregonensis.* ADULT: Moderately small (BL 9–15 mm); dull metallic blue; wing covers with rows of fine indentations. Males resemble females but have greatly elongated mandibles. LARVA: White, C-shaped grubs with a darker head; commonly found in rotten wood of California Bay trees *(Umbellularia)*. RANGE: Santa Cruz County northward and Sierra Nevada.

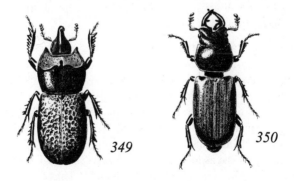

Figs. 349–350 Stag Beetles: 349, Rugose Stag Beetle, *Sinodendron rugosum;* 350, Oregon Stag Beetle, *Platycerus oregonensis.*

SCARABS
Family Scarabaeidae

Scarabaeids are small to large, robust, dull or brightly colored beetles with antennae that terminate in 3–7 broad, leaflike segments which can be closed together to form a tight club. The family includes a tremendous diversity of forms and behavioral adaptations: larvae of many live among roots or in rotting plant material, especially rotting wood, others in dung, ant or termite nests, burrows of vertebrates, carrion, or fungus. Adults often feed on living plants and damage agricultural crops independently from the larvae. They include some of the most notorious of all insects, such as the Japanese Beetle.

The larvae are C-shaped grubs commonly with about 3 fleshy lobes to each segment, except for the last 3 which are usually smooth, shiny, and semi-transparent. The legs lack the ridges used in making rasping noises in the Lucanidae.

This is a large, worldwide family, well represented in Temperate and Tropical zones, with upwards of 500 species in California.

351. European Dung Beetle. *Aphodius fimetarius*. ADULT: Small (BL 5–7 mm); oblong, convex; black with bright, reddish brown wing covers. LARVA: Head brown, body whitish; in cow pats and other dung. RANGE: Widely distributed in California wherever animals are grazed, except in arid areas; introduced from the Old World.

Plate 14b. Rain beetles. *Pleocoma*. With the first soaking rains in fall, males of this genus suddenly appear in abundance, flying at dusk or on cloudy, drizzly days in search of the flightless females. Evidently females give off a scent that attracts males to their burrows, where mating occurs. After mating, the females plug the tunnels and deposit eggs. The larvae feed on roots of hardwood trees and shrubs, and the life cycle is reported to last as long as 10 or 12 years. ADULTS: Moderately large (BL 20–40 mm); the females larger and more robust; wing covers shiny, black, reddish brown, or nearly translucent; underside clothed with dense hair. The front of the head is broadened, bilobed, and flattened, an adaptation for

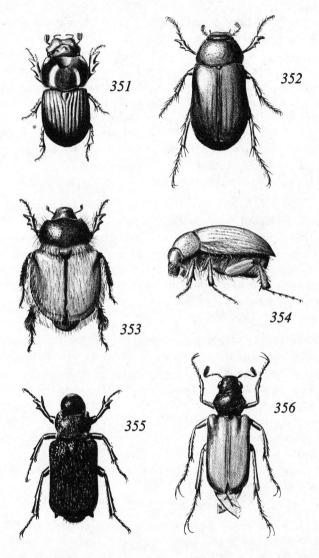

Figs. 351–356 Scarab Beetles: 351, European Dung Beetle, *Aphodius fimetarius;* 352, Common June Beetle, *Phyllophaga* species; 353, Little Bear, *Paracotalpa ursina;* 354, Serica, *Serica* species; 355, Ant Scarab, *Cremastocheilus* species; 356, Green Pine Chafer, *Dichelonyx backi.*

scooping soil. LARVAE: Head amber colored; body shining white with tiny, raised black spots peppering the back. RANGE: Throughout the foothills and mountainous parts of the state. The ranges of most of the species are restricted, and isolated, like islands. In southern California, common species include the **Black Rain Beetle** *(P. puncticollis)* in the Santa Monica Mountains, which is shining black with black hair, and the **Southern Rain Beetle** *(P. australis)* in the San Gabriel Mountains, a black species with reddish brown hair.* In the north, **Behren's Rain Beetle** *(P. behrensi)* in the Berkeley Hills, which is dark mahogany brown in the male, reddish brown in the female, with pale hair, is the most commonly collected species.

352. Common June beetles. *Phyllophaga.* ADULTS: Moderately large (BL 13–24 mm); dark to light brown; smooth, usually dusted with scales. They often occur in large numbers at lights on warm summer evenings. LARVAE ("white grubs"): C-shaped with a dark head and pale, deeply wrinkled body; on roots of various plants. RANGE: Sierra Nevada southward, including the deserts.

Plate 14c. Ten-lined June Beetle. *Polyphylla decimlineata.* ADULT: Large (BL 22–35 mm); robust; brown, with white scales that form longitudinal stripes on the head, thorax, and wing covers and with scattered yellowish scales between the stripes; antennae reddish brown in male, with very large leaves forming the club. LARVA: Large (BL 25–50 mm); whitish C-shaped grub. On roots of shrubs and trees including ornamental shrubs and nursery trees; 2 or 3 years may be required to complete growth. The beetles appear in spring and early summer and are seen flying at dusk low over fields. RANGE: Throughout most of the state at low to moderate elevations.

353. Little Bear. *Paracotalpa ursina.* ADULT: BL 10–23 mm; robust, round in outline; steel bluish with bright reddish brown or black wing covers; body clothed with long, yellowish hairs. RANGE: Lower foothills of southern California and margins of the Central Valley. The beetles appear in large numbers in spring, flying low over grassy fields during the day. In mountains of southern California the beetles are black or metallic green with black wing covers.

Plate 14d. Green Fruit Beetle. *Cotinus texana.* ADULT: Large (BL 20–34 mm); robust, somewhat pentagonal in outline, widest at the base of the wings; body bright metallic green; back and wings velvety olive green with marginal bands of brownish yellow. The beetles feed on a wide variety of fruits including cacti, figs, peaches, and grapes. LARVA: Burrow in and under decaying or dry horse dung on which it feeds. RANGE: Southern California, where it has appeared in recent years, having expanded its range from Arizona; common near stables.

354. Sericas. *Serica.* ADULTS: Medium-sized (BL 7–11 mm); cylindrical; smooth or somewhat velvety, black, brown, or reddish brown with faintly lined wing covers. They sometimes appear in great numbers at lights or on flowering shrubs and trees in spring and may cause damage by defoliating fruit trees. RANGE: Throughout most of the state at a wide range of elevations, especially in foothill and mountainous areas.

355. Ant scarabs. *Cremastocheilus.* ADULTS: Medium-sized (BL 10–16 mm); rather flat with pentagonal outline; usually black with faint, whitish markings. They are most often seen solitary, flying or on the ground, but occasionally under bark or rocks in association with ant colonies. They are predaceous on ant larvae. LARVAE: Covered with long, silky hairs; in nests of mound-building harvester ants, where they live as scavengers. RANGE: Widely distributed from foothills to middle elevations of the Coast Ranges and Sierra Nevada.

356. Green chafers. *Dichelonyx.* ADULTS: BL 7–12 mm; rather slender with the thorax narrower than the abdomen; reddish brown or black with metallic green wing covers. RANGE: Throughout chaparral and forested parts of California. **(356) Green Pine Chafer** *(D. backi)* is variable, reddish brown to reddish black with bright green to brown or greenish black wing covers. It is common on pine, fir, and Douglas Fir throughout the Sierra Nevada and Cascade Range.

WATER PENNIES
Family Psephenidae

Water pennies receive their common name from the remarkable larvae, which are round or ovoid in outline, extremely flattened, almost scalelike. The margins of the body are greatly expanded and cover the head and legs, enabling the larvae to cling to rocks in rapids of streams. They breath by means of abdominal gills or a retractile tuft of anal filaments. The adults are terrestrial, although females may enter the water to lay eggs. Psephenid adults are small, oval, flattened beetles with long, feathery antennae in males; found on streamside rocks or vegetation, but not commonly encountered because adults are present only a few weeks each year. Although larval water pennies are quite common and characteristic inhabitants of moderate to rapid streams throughout California, only 6 species are recorded in the state.

357. Edwards' Water Penny. *Eubrianax edwardsi.* ADULT: Small (BL 3–6 mm); dark brown or black (female) or with wing covers pale tan (male); head hidden beneath expanded thorax. LARVA (**357**): Broad, flat, round-oval; with the segments fitted tightly together to the outer margin; abdomen with lateral gills on segments 1–4; on rocks in well-aerated water which is protected from erosion and silting. RANGE: Streams throughout California below 2000 m (6500 ft). Members of the related genus, *Psephenus,* are similar, but in the adult (**358**) the head is not hidden beneath the pronotum. The larva has gills on segments 1–6 or 2–6.

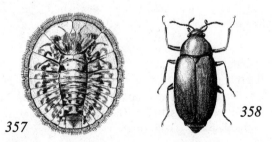

Figs. 357–358 Water Pennies: 357, Edwards' Water Penny, *Eubrianax edwardsi* (larva); 358, Water Penny, *Psephenus* species (adult).

Family Buprestidae

Buprestids are small to moderately large beetles, including many of beautiful metallic colors. They are generally streamlined in appearance, elongate or oval and flattened, often etched with longitudinal grooves. The larvae bore in roots, branches of woody plants, or under bark of fallen logs, and a few species are leaf or stem miners. They are elongate, flat, strongly segmented grubs with a small head almost entirely withdrawn into the greatly enlarged first thoracic segment. This characteristic body form and habit give rise to the somewhat erroneous, yet descriptive common name, "flat-headed borers."

The adults of some genera are most often encountered at flowers, where they congregate to feed on pollen; others are found on recently cut trees or felled branches. Many are exceedingly fast fliers and take off at the slightest alarm. More than 200 species occur in California.

Plate 14e. Spotted flower buprestids. *Acmaeodera*. Members of this large genus are common on a variety of flowers in the foothills and deserts. ADULTS: Small (BL 4–14 mm); somewhat flattened, bullet-shaped, broadest at the base of the wings; tan, brown, or black with yellow, orange, or red markings on the wing covers, and covered with soft hair. LARVAE: White with a brown head; in dead or dying branches of various shrubs and broadleafed trees. RANGE: Throughout low- to middle-elevation parts of the state, most common in oak woodland and desert areas.

359. Sculptured Pine Borer. *Chalcophora angulicollis*. This beetle is common on recently felled conifer logs in lumbering areas. ADULT: Large (BL 24–32 mm); rather broad with wing covers slightly wider than the thorax; shining dark gray or black with irregular, sculptured, bronzy areas on the back; underside metallic bronze colored. LARVA: Large, white, flattened grub; under bark of pine, fir, and Douglas Fir. RANGE: Throughout coniferous forests of California.

Plate 14f. Golden Buprestid. *Buprestis aurulenta*. This is one of California's most beautiful beetles and is one of our commonest mountain buprestids. The adults sometimes congregate at the new tips of young pines where they apparently feed on exudates. ADULT: Moderately large (BL 14–20 mm); bright iridescent green with the median line and lateral margins of the wing covers golden coppery or reddish; with alternating longitudinal grooves and ridges on the wing covers. LARVA: Large, pale, flat-headed borer in injured or dying Douglas Fir and other coniferous trees, especially those scorched in fires. RANGE: Throughout coniferous forests of California.

360. Anthaxias. *Anthaxia.* ADULTS: Small (BL 3.5–7 mm); bright metallic green, or more commonly black with a dull metallic green underside. Often abundant on flowers. LARVAE: Elongate, thin, white with moderately expanded thorax;

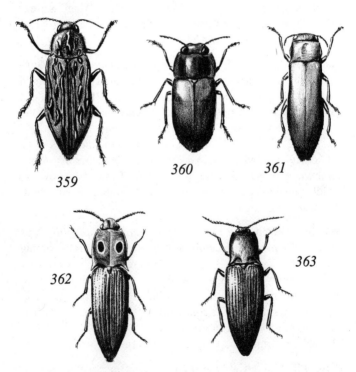

359

360

361

362

363

Figs. 359–363 Beetles: 359, Sculptured Pine Borer, *Chalcophora angulicollis;* 360, Anthaxia, *Anthaxia* species; 361, Twig Girdler, *Agrilus* species; 362, Eyed Elater, *Alaus melanops;* 363, Common Click Beetle, *Limonius* species.

in bark or outer wood of living or dying trees including conifers and shrubs. RANGE: Throughout chaparral and forested
parts of the state; most common in oak and mixed oak-conifer
communities.

361. Twig girdlers. *Agrilus*. Members of this genus are slender compared to other buprestids and are often collected by
beating or sweeping host plants such as willow and oak.
ADULTS: Mostly small (BL 3–14 mm); slender with the head,
thorax, and wing covers of about equal width; most often dark
metallic bronze, purplish, or greenish. LARVAE: Very elongate (BL up to 30 mm); in spiral galleries around twigs of
broadleafed trees and shrubs. RANGE: Throughout lowland
riverine, foothill, and middle-elevation areas of the state.

CLICK BEETLES
Family Elateridae

Click beetles are rather slender, somewhat flattened, and
characteristically shaped, with the thorax rectangular in outline, drawn out to pointed posterior corners, and with narrow,
oval wing covers. Elaterids also are easily recognized by a
posteriorly directed process on the underside of the first
thoracic segment, and a corresponding groove on the second
segment. By articulation of these 2 segments the process is
snapped into the cavity, resulting in an audible "click". When
the beetle is lying on its back, this enables it to spring several
cm into the air. Most species are brown or black, and some are
nocturnal and come to lights.

The larvae ("wireworms") are elongate, thin, and cylindrical with heavily armored skin. They live in the soil or rotten
wood and are either vegetarians, eat both plant and animal
material, or are predaceous. Some species are of economic
importance, destructive to the roots of field crops. This is a
large family, with more than 250 species known in California.

362. Eyed Elater. *Alaus melanops*. This is the largest elaterid
in California; it is most often found under loose bark of old
conifer stumps. ADULT: Large (BL 22–40 mm); black with
variable whitish scaling and 2 conspicuous, oval, eyelike spots
on the prothorax which are velvety black and often surrounded
by rings of whitish scales. LARVA: BL 50–60 mm; smooth,

cylindrical, yellowish with the head and thorax dark brown; in rotten wood of logs and stumps; predaceous on larvae of wood-boring beetles. RANGE: All coniferous forest regions of California.

363. Common click beetles. *Limonius, Ctenicera.* Members of these genera are small beetles, most often seen at flowers where they feed on petals and foliage. ADULTS: *Limonius,* with about 20 species, are small (BL 4–13 mm); slender, dark brown, in some species with the "shoulders" or most of the wing covers red-brown. *Ctenicera,* represented by more than 50 species, are larger (BL commonly 6–20 mm, rarely to 26 mm); relatively broader, the thorax and wing covers oval in outline; often black, sometimes with red markings on the thorax, or have tan wing covers marked with transverse bands or zig-zag markings. LARVAE: Thin; brown to amber-colored wireworms; with a notch in the hind margin of the last body segment and one or more spurs or projections on each side of the notch. They occur normally in soil, decaying wood, or forest litter, occasionally in other habitats including old cow dung and fungus. They live for 3 or more years and are opportunists, eating living roots, decaying organic matter, or other larvae. RANGE: Throughout most of the state in wooded and agricultural areas.

GLOWWORMS
Family Phengodidae

Glowworms are remarkable because the females are larviform, much larger than the males, and have rows of luminescent spots (sometimes of 2 colors, although not in California species). They have immense powers of attraction to the males by means of a chemical (pheromone). Males resemble leather-winged beetles but have feathery antennae and shorter elytra than most cantharids. They fly in the evening and sometimes come to lights. The larvae, which also have luminescent bands or spots, feed on millipedes and other invertebrates. The family is small and restricted to the Americas, with 6 species in California.

364; Plate 14g. California Phengodid. *Zarhipis integripennis.* ADULT: Male (**364**) BL 10–20 mm; orange with a black or orange head and blue-black wing covers slightly shorter than the abdomen; female (**Plate 14g**) larviform and much larger (BL 30–80 mm); amber or buff colored with black markings on the back, and spots along the sides that luminesce bright green; active at night. LARVA: Resembles the female but smaller. Females and larvae subdue millipedes by curling around them snake-fashion and paralyzing them by biting them ventrally just behind the head. The glowworm enters just behind the millipede's head and eats its way tailward, leaving only the segmental rings of the shell. RANGE: Coast and Transverse ranges and Sierra Nevada, foothills to middle elevations.

FIREFLIES
Family Lampyridae

The family Lampyridae includes some of the most familiar of all insects in humid regions, including the eastern United States, but in California we have only a few species and none is

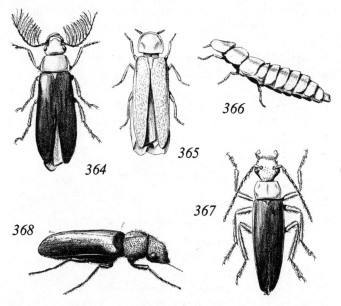

Figs. 364–368 Beetles: 364, California Phengodid, *Zarhipis integripennis* (male); 365, Pink Glowworm, *Microphotus angustus* (male); 366, *M. angustus* (female); 367, Downy Leather-Winged Beetle, *Podabrus pruinosus;* 368, Stout's Branch Borer, *Polycaon stouti.*

luminescent as a winged adult. The lights are borne on the hind segments of the abdomen in most lampyrids; in our species these are lacking in males, and present only in the females of certain species which are wingless ("glowworms"). The larvae are predaceous on snails and slugs, and sometimes other invertebrates.

365, 366. Pink Glowworm. *Microphotus angustus.* ADULT: Male (**365**) BL 8–10 mm; with grayish brown prothorax and elytra and a pinkish body; female (**366**) BL 10–15 mm; wingless, larviform, flattened; pinkish; luminesces with a bright green light. They occur in dry grassy areas in summer and late spring. RANGE: Foothills of the Coast Ranges and Sierra Nevada. A larger and more commonly seen species, especially in northern California, is the **California Glowworm** *(Ellychnia californica).* ADULT: Medium-sized (BL 8–16 mm to tips of wings); black with broad, rounded wing covers and thoracic shield which has 2 bright rosy red marks near the lateral margins. They are often found on low, grassy vegetation during the daytime. LARVA: Undescribed; larvae believed to be this species by circumstantial association have bright, pale green luminescent areas on the abdomen; black plates on the back separated by pale bands between the segments; BL 15–18 mm. They are active at night during winter, feeding in native land snails, and the lights can be seen at a distance, even through the snail shell. RANGE: Wet areas of the Coast Ranges.

SOLDIER or LEATHER-WINGED BEETLES
Family Cantharidae

Cantharids are most easily distinguished by their narrow-rectangular shape, soft wing covers, and large thoracic shield which is expanded but does not cover the head. They have long, threadlike antennae and well-developed wings in both sexes. Adults may be general feeders or feed mostly on plants, and larvae are predaceous. The beetles are most often seen on vegetation in the daytime, although some species come to lights. There are over 100 species in California; the great majority of them are similar in appearance, red or orange with gray, black, or brown wing covers.

Plate 14h. Brown Leatherwing. *Cantharis consors.*
ADULT: Moderately large (BL 14–19 mm); body orange,
wing covers brown, and legs reddish marked with black.
Common in a variety of habitats, often found on vegetation and
at lights, but the larva is unknown. RANGE: Californian province; Coast Ranges and coastal valleys of central and southern
California.

367. Downy Leather-winged Beetle. *Podabrus pruinosus.*
ADULT: Small (BL 9–15 mm); head and body pale orange,
wing covers blackish to pale brown, with a covering of short,
whitish hair giving a grayish blue appearance. Common in
spring and summer eating aphids on flowering shrubs. They
feign death by curling the head downwards and dropping to the
ground when disturbed. LARVA: BL 15–20 mm; pink, covered with fine hair, and with 2 longitudinal dark lines on the
back of the thorax; in soil. RANGE: Coastal valleys to middle
elevations of the Coast Ranges and Sierra Nevada, especially
in chaparral areas.

BRANCH AND TWIG BORERS
Family Bostrichidae

Members of the family Bostrichidae make cylindrical burrows in dried wood, especially hardwoods, including dead
limbs and fire-killed trees, and sometimes cause damage by
mining limbs of living cultivated trees. The beetles characteristically are cylindrical with the head turned downward. Often
the thorax and elytra are heavily sculptured and the body is
truncated at the tail end. The larvae are somewhat scarablike,
curved, and resemble those of anobiids but have a smaller head
and more enlarged thoracic region. About 20 species occur in
the state, and several others are often intercepted in quarantine
inspections at ports.

368. Stout's Branch Borer. *Polycaon stouti.* ADULT: One
of the largest bostrichids in California but greatly variable in
size (BL 10–23 mm); cylindrical, black or reddish black; hairy
but not strongly sculptured; head only slightly turned down.
LARVA: Burrows in the heartwood of dead hardwood trees.
RANGE: Coastal valleys and mountains to middle elevations.

The **California Branch Borer** *(P. confertus)* is smaller (BL 7–14 mm); generally dark reddish brown. It has the same habits, and the adults sometimes burrow into living orchard trees and cause damage. RANGE: Widespread in foothill and agricultural areas. Another California bostrichid is called the **Lead Cable Borer** *(Scobicia declivis)* because it sometimes bores holes in lead telephone cables through the lower side where the hangers support the cables. Short-circuits ultimately result. The beetles are small (BL 3.5–7 mm), black, with the head more strongly turned down than in *Polycaon*.

HIDE BEETLES
Family Dermestidae

Dermestids are small, compact, oval beetles that are covered with fine hair or appressed scales. The sides of the body are grooved for reception of the antennae and legs, which are drawn in tightly when the beetle is alarmed. The larvae are elongate, cylindrical, somewhat caterpillarlike, with bands of stiff, jointed hairs. They are scavengers on dried animal and plant material of high protein content, including furs, hides, wool, feathers, debris in spider webs and wasp nests, and stored grain. Adults of some dermestids are common flower visitors where they feed on pollen. About 40 species are known in California.

Plate 15a. Carpet beetles. *Anthrenus.* A menace to every insect collector, the larvae of this genus seem to appear out of nowhere whenever a collection is left unattended and to go unerringly to the most prized specimens. If not deterred by preventative chemicals such as naphthalene flakes, they can reduce an entire collection to dust in a few months. They wreak similar havoc in stored woolen fabrics, feathered hats, preserved animal skins, etc. ADULTS: Small (BL 3–4 mm); oval, convex; black, covered with white, brown, orange, and yellow scales that form a checkered pattern. They are often found on window sills in infested buildings and in nature at flowers. LARVAE: Pale brownish with narrow bands of long, stiff, brown hairs and longer tufts at the tail end; in wool, museum specimens, and other animal products. RANGE: Throughout most of the state except the deserts.

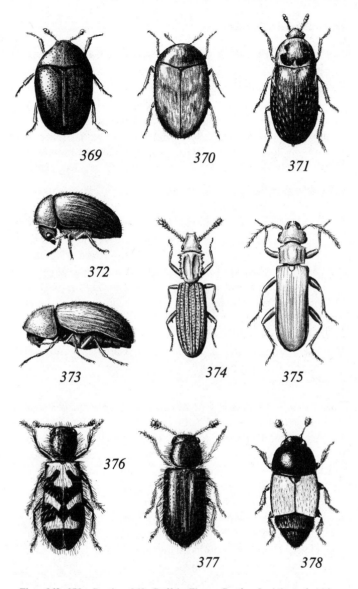

Figs. 369–378 Beetles: 369, Buffalo Flower Beetle, *Orphilus subnitidus;* 370, Khapra Beetle, *Trogoderma granarium;* 371, Common Carrion Dermestid, *Dermestes marmoratus;* 372, Cigarette Beetle, *Lasioderma serricorne;* 373, Drugstore Beetle, *Stegobium paniceum;* 374, Saw-Toothed Grain Beetle, *Oryzaephilus surinamensis;* 375, Red Flat Bark Beetle, *Cucujus clavipes;* 376, Common Checkered Beetle, *Trichodes ornatus;* 377, Red-Legged Ham Beetle, *Necrobius rufipes;* 378, Cactus Flower Beetle, *Carpophilus pallipennis.*

369. Buffalo Flower Beetle. *Orphilus subnitidus.* ADULT: Similar in form to carpet beetles but almost pentagonal in outline and slightly larger (BL 3–5 mm); black without any scale patterns. They are often abundant on flowers such as yarrow *(Achillea)* and Cow Parsnip *(Heracleum lanatum)*. LARVA: Apparently not described; probably in fungus or fungus-infested wood. RANGE: Coastal areas and mountains to middle elevations. Common in spring and early summer.

370. Khapra Beetle. *Trogoderma granarium.* This is one of the worst stored-products pests in the world. Introduced into California about 1946, it spread in granaries through the Imperial and Central valleys during the following several years and was subjected to intensive survey and fumigation programs. ADULT: Small (BL 3–4 mm); oval, brownish black; does not feed and seldom flies, dispersal occurring through movement of stored products. LARVA: BL to 6 mm; brownish with yellowish intersegmental rings and bands of short and long hairs; in grains, flour, cereal products, dried fruits, nuts, etc. RANGE: Believed to be eradicated in California, although similar species of *Trogoderma* occur widely in urban and agricultural areas.

371. Common Carrion Dermestid. *Dermestes marmoratus.* The largest dermestid in California (BL 9–14 mm), this species is common around animal carcasses in later stages of decomposition. ADULT: Elongate-oval, slightly flattened; black with gray scales forming spots and a broad band across the base of the wing covers; underside mostly whitish. LARVA: Elongate, reddish brown with a pale stripe down the back, and covered with long, reddish brown hairs; on hides, bones, etc. RANGE: Throughout the state at low to middle elevations. Probably introduced by hide traders in the early days of commerce. Two similar species also are common in California on carcasses. The **Hide Beetle** *(D. maculatus)* has been used successfully for many years to clean skeletons of vertebrate specimens for museums. ADULT: Smaller (BL 7–10 mm); generally brownish black, lacking the whitish spotting on the upperside; distinguished by a spine and saw-toothlike edge on the hind margin of each wing cover.

RANGE: Widespread at lower elevations especially in shaded situations. **Frisch's Carrion Beetle** *(D. frischi)* is similar in size and color but lacks the spine and toothed edge on the elytra.

STORE AND FURNITURE BEETLES
Family Anobiidae

Anobiids are small, mostly brown beetles with an oval or elongate form and characterized by having the head somewhat withdrawn under the hoodlike prothorax. When the insect is startled, the legs and antennae can be drawn up into sockets, and it feigns death. Most members of the family feed in wood (dead branches, woody galls, etc.) in nature, and sometimes they are destructive to furniture and flooring, riddling them with small, round emergence holes. A few species, including the 2 illustrated, are stored-products pests. The larvae are crescent-shaped with enlarged thoracic area and tail end, resembling those of scarab beetles but with the legs much reduced or absent. About 75 species are known in California.

Sometimes anobiids are called "death-watch beetles." This name originates from British folklore and suggests ghost stories set in old castles. The beetles create a noise by bumping their heads on the top of the tunnels as a telegraphic message at mating time. An eerie tapping is audible in silent rooms, such as during a bedside death-watch.

372. Cigarette Beetle. *Lasioderma serricorne.* This species has similar habits, foods, and geographical range to the following species. It is particularly well known because of its damage of cured tobacco, in leaf and in cigarettes and cigars. ADULT: Small (BL 2–3 mm); round, dark brown; clothed with short, fine hair and without striated elytra.

373. Drugstore Beetle. *Stegobium paniceum.* ADULT: Small (BL 2.4–4.0 mm); slender; elytra reddish brown with narrow, parallel grooves; body clothed with dense, short hairs. LARVA: Small; white; C-shaped; clothed with long hairs. The species occurs around the world. Larva feeds in all kinds of cereals, dried fruits, nuts, etc., and in dried animal products including wool, hair, leather, and plant and insect museum specimens. It is often encountered in packaged paprika. RANGE: Urban and rural areas inhabited by man.

Family Ostomatidae

Plate 15b. Green Ostomatid. *Temnochila chloridia.* ADULT: Moderately large (BL 9–7 mm); slender, nearly cylindrical; metallic green or blue. LARVA: Slender, somewhat flattened; white or pink with shining black or dark brown head, thoracic shield, and anal plate bearing 2 short, posteriorly directed spurs. Both adults and larvae are predaceous, foraging under bark of fallen trees, in galleries of wood-boring insects, or in wood-rot fungi. RANGE: Throughout forested portions of the state, including the desert mountains.

FLAT BARK BEETLES
Family Cucujidae

Typical members of the family Cucujidae are rather elongate, flattened, adapted for life under bark. Other cucujids live in decaying plant material, dry fruit, and cereals, and a few are pests of stored products. Less than 30 species occur in California, many of which are very small.

374. Saw-toothed Grain Beetle. *Oryzaephilus surinamensis.* This species lives in all kinds of stored foods and has been carried by commerce to all parts of the world. ADULT: Small (BL 3–4 mm); reddish or dark brown; slender, with ridged wing covers and 6 saw-teeth on each side of the thorax. LARVA: Slender; the body tapering to a posterior tip; white with brown head and yellow plates on each segment; in cereal, dried fruits, sugar, tobacco, etc. RANGE: Urban and agricultural areas throughout the state and out of doors in decayed, dry vegetable products.

375. Red Flat Bark Beetle. *Cucujus clavipes.* This is the largest and most conspicuous cucujid in California. ADULT: BL 12–17 mm; bright red with black eyes, antennae, and apical leg segments; somewhat rectangular in outline, very flattened, with rows of fine indentations on the wing covers. LARVA: Also very flattened, amber colored, elongate with head wider than body segments, the last 2 body segments equal in length, and there is a 2-pronged process on the tail end; lives under bark of logs and stumps. RANGE: Throughout coniferous forests of the state. The larvae may be confused with those of another beetle, family Pyrochroidae, which live under bark and are similar in appearance but have the last body segment before the pronged tail plate about twice as large as the preceding segment.

CHECKERED BEETLES
Family Cleridae

Clerids are slender, hairy, often brightly colored beetles. Most are found at flowers or elsewhere on plants, although a few occur in animal carcasses or on recently felled trees. The larvae are nearly all predaceous on wood-boring Coleoptera and other insects, such as in cones and seeds, galls, foliage, and dead twigs. There are nearly 100 species in California, inhabiting a wide range of habitats from conifer forests to the deserts.

376. Common Checkered Beetle. *Trichodes ornatus.* This species is found on flowers in many different ecological situations. ADULT: BL 5–13 mm; metallic bluish with variable amounts of irregular, bright yellow or orange banding on the wing covers. LARVA: BL to 13 mm; slender, yellowish with head, thoracic shield, and anal plate dark brown; feeds on pollen and animal contents of bees' and wasps' burrows. RANGE: Throughout most of California, foothills to timberline.

377. Red-legged Ham Beetle. *Necrobius rufipes.* ADULT: Small (BL 4–5 mm); green or blue with the bases of the antennae and legs reddish. LARVA: Yellowish or brownish with the head and thorax darker brown; somewhat hairy. Adults and larvae are predaceous on fly maggots and other insects in decaying meat. The larvae may also feed on stored foods of animal and vegetable origin. RANGE: Widespread in low-elevation areas and occasional at high elevations in arid parts of the state.

SAP BEETLES, DRIED FRUIT BEETLES
Family Nitidulidae

Nitidulids are variable in form but most often are rather broad (with the thorax as wide as the elytra), somewhat flattened beetles with clubbed antennae and short wing covers which expose the tip of the abdomen. Many species are found in flowers, others live in wood-rot fungi, decaying fruit, or animal matter. The larvae of most burrow into plant substances. About 50 species are known in California.

378. Cactus Flower Beetle. *Carpophilus pallipennis.* ADULT: Small (BL 3–4 mm); brown or reddish brown with the wing covers pale amber or yellowish. Common in flowers, especially those of cacti. LARVA: Yellowish, blunt, cigar-shaped; with short legs; sparse hairs; and two short, dark spurs at the tail end; in old cactus fruit and other decaying vegetable matter, rarely in stored foods. RANGE: Desert areas and other arid parts of southern California. Other species of *Carpophilus* are common around decaying fruit, wherever fruit is grown in the state.

BLISTER BEETLES
Family Meloidae

The Meloidae is one of the most interesting of all Coleoptera families because of the remarkable life history. The beetles are variable in form but have the head turned down and generally are soft bodied with soft, leathery wing covers, which in some species are shorter than the abdomen. Females deposit immense numbers of eggs, usually on the ground. The larvae go through successive stages which differ in form: at first they are active, leathery skinned, and look like minute silverfish (triungulins). At this stage they wander in search of hosts (specific kinds of Hymenoptera or egg cases of Orthoptera) and most perish. Those that locate a host, molt into maggotlike, short-legged forms. Finally, in many species, stages of pupalike larvae and further feeding instars complete the larval growth. The adults eat plant material, especially flowers.

The "Spanish Fly," *Lytta vesicatoria,* of southern Europe yields the pharmaceutical product cantharides, which is prepared from dried beetles and was formerly used as an aphrodisiac. Many other species, especially in tropical regions, secrete defensive fluids (cantharidin) that cause severe blistering of human skin. Few if any of the 100-plus species in California possess this capability, despite the common name.

Plate 15c. Green Blister Beetle. *Lytta chloris.* ADULT: Medium-sized (BL 7–14 mm), slender; bright metallic green or bluish. Common on spring wildflowers, especially lupine and other members of the pea family. LARVA: In nests of ground-nesting bees. RANGE: Foothills of the Coast Ranges and Sierra Nevada. The **Small Black Blister Beetle** *(L. nigripilis),* which has a similar distribution, is about the same size and entirely black. It is common in late spring and early summer.

379. Punctate Blister Beetle. *Epicauta puncticollis.* This is perhaps California's most common meloid outside of the desert

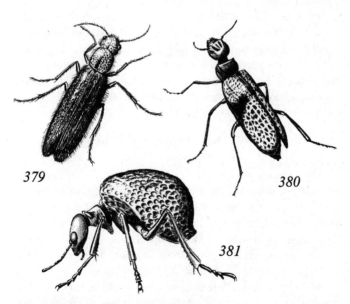

Figs. 379–381 Blister Beetles: 379, Punctate Blister Beetle, *Epicauta puncticollis;* 380, Soldier Blister Beetle, *Tegrodera* species; 381, Desert Spider Beetle, *Cysteodemis armatus.*

areas. ADULT: Medium-sized (BL 7–15 mm); black with a bluish sheen; entire surface closely punctured with minute dents. Most commonly seen in late spring and summer on tarweed *(Hemizonia)* and related plants, but the beetles sometimes eat flowers of various field crops, especially in late summer. LARVA: On grasshopper egg masses. RANGE: Coastal areas, foothills, and mountains to middle elevations all over the state.

380. Soldier blister beetles. *Tegrodera.* ADULTS: Large (BL 15–30 mm); elongate with somewhat broadened abdomen and wing covers; head red, thorax reddish, body and elytra black with a network of raised, golden yellow lines covering all but a belt in the middle and the posterior tips. The beetles feed on a wide variety of wildflowers. RANGE: There are 3 species with varying amounts of yellow on the elytra, 1 occurring in the Owens Valley and 2 in the low desert, Imperial Valley, and inner coastal valleys of southern California.

381. Desert Spider Beetle. *Cysteodemis armatus.* ADULT: This flightless beetle is spiderlike, with inflated abdomen, small head and thorax, and long legs; BL 7–17 mm; black, blotched with variable white or yellow, waxy secretion. The wing covers are not as soft as in most meloids and are fused, forming a heavily sculptured shell over the abdomen. RANGE: Desert areas.

DARKLING GROUND BEETLES
Family Tenebrionidae

Tenebrionids are abundant and diverse, with more than 400 species in California. The larvae are found in almost all terrestrial habitats: in rotten wood, wood-rot fungi, sand dunes and other soil, in termite and ant nests, stored-food products, etc. The adults occur with the larvae, under bark and debris on the ground, or are seen running about in desert habitats or other arid places, especially at dusk.

Although nearly all Temperate Zone tenebrionid beetles are black or dark brown, they exhibit a wide range of dissimilarity in form, being tiny to large, slender or robust, flattened or convex. Some are quite active and resemble predaceous ground beetles, while many are sluggish, and often they are flightless, with vestigial hindwings and fused wing covers. The larvae (''false wireworms'') are mostly uniform in character, elongate, rather slender, thick skinned, amber colored or reddish brown, usually without conspicuous spurs or other armature.

382. Live Oak Cluster Beetle. *Cibdelis blaschkei.* ADULT: BL 13–15 mm; stout, oval, slightly flattened; black with the thorax smooth, dull, the wing covers faintly roughened. Often found in aggregations under loose bark of oak in winter and early spring. RANGE: Coast Ranges and Sierra Nevada.

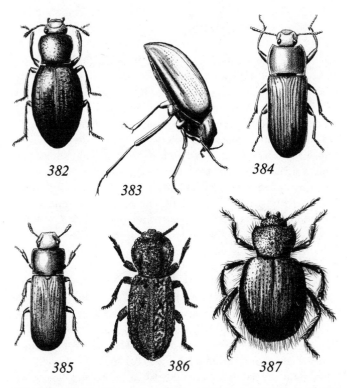

Figs. 382–387 Darkling Ground Beetles: 382, Live Oak Cluster Beetle, *Cibdelis blaschkei;* 383, Armored Stink Beetle, *Eleodes armata;* 384, Yellow Mealworm, *Tenebrio molitor;* 385, Flour Beetle, *Tribolium* species; 386, Ironclad Beetle, *Phloeodes diabolicus;* 387, Woolly Ground Beetle, *Cratidus osculans.*

383. Stink beetles or **olive beetles.** *Eleodes*. Members of this genus are legion, with about 100 species represented in California. They are medium-sized to large, robust, usually smooth, black beetles. They have the interesting habit of standing on their heads and spraying or oozing an offensive-smelling vapor as protection against predators. ADULTS: BL 8–40 mm; robust with the wing covers tapered or attenuated to the posterior tip; with long legs. Often seen wandering over sandy terrain. LARVAE: Slender, nearly cylindrical; amber to brown, with the last abdominal segment pointed and upcurved; in sand or soil, feeding on roots. RANGE: Throughout the state. The **(383) Armored Stink Beetle** *(E. armata)* is large (BL 17–33 mm); widespread in arid areas, coastal southern California, deserts, and Central Valley. It has prominent spurs on all the legs. The **Dentate Stink Beetle** *(E. dentipes)* is smaller (BL 16–28 mm); variable in shape, with the tooth on the front leg of both sexes (which also occurs in several other species). It ranges from San Luis Obispo County northward and is common in stumps, logs, dry leaves, etc. The 5 or 6 species of another genus, *Coelocnemis,* are similar to *Eleodes* but have lines of golden hairs on the inner margins of the legs (tibiae and tarsi). All of these beetles produce quinonoid secretions which are believed to make them unpalatable to would-be vertebrate predators, such as rodents. Although *Eleodes* and *Coelocnemis* are not considered to be closely related on evolutionary bases, pairs of species of the two that live together mimic one another in dorsal appearance, thus presumably mutually aiding each other in defense against predators.

384. Yellow Mealworm. *Tenebrio molitor*. This cosmopolitan species lives in stored cereals and is often reared in colonies for experimental purposes or as food for lab animals or pets. ADULT: Medium-sized (BL 13–15 mm); elongate-oval in outline, somewhat flattened; brown or blackish with longitudinal lines on the wing covers. LARVA: BL to 25 mm; slender, convex above, flat beneath; yellowish or brownish with darker bands at the segmental joints; in grains, seeds, etc. RANGE: Throughout urban and agricultural areas of the state and occasionally out of doors at lower elevations. The larvae are sold as fish bait ("Li-cut worms").

385. Flour beetles. *Tribolium.* ADULTS: Small (BL 3–4 mm); slender, flattened, with smooth margins on the thorax; pale to dark reddish brown, with the head and thorax darkest. LARVAE: Nearly cylindrical, strongly segmented; white to yellowish with dark mandibles and a pair of short tail spurs; adults and larvae commonly found together in all kinds of cereal products, nuts, and other dried foods. RANGE: Urban and agricultural areas throughout the state.

386. Ironclad Beetle. *Phloeodes diabolicus.* This species is sometimes placed in another family, Zopheridae, because the larvae are very different from other tenebrionids. This beetle is named for its extremely hard shell. Specimens have the peculiar ability to postpone breathing and survive long periods in killing jars with cyanide or even immersed in alcohol, and when finally killed are difficult to pierce with a pin. ADULT: BL 15–23 mm; oval in outline with the thorax as broad as the abdomen, flattened and rough-surfaced; dark brown, mottled with black and gray. Common under loose bark of dead oak and cottonwood. LARVA: Superficially like a flat-headed borer (Buprestidae), with enlarged thorax and reduced legs; in rotten wood. RANGE: Woodlands throughout much of California.

387. Woolly Ground Beetle. *Cratidus osculans.* ADULT: BL 12–16 mm; rectangulate, robust; black; sparsely covered with long, reddish brown hair. Often seen along roads or paths in wooded or chaparral areas. RANGE: Coastal, southern to central California.

LADYBIRD BEETLES
Family Coccinellidae

Coccinellids are uniformly oval or round beetles that are strongly convex above, flat beneath, shining, and often brightly colored. "Ladybugs" are among the most familiar of all insects, and their beneficial effects to gardeners have long been known. Both adults and larvae of most species eat aphids, mealybugs, and similar soft-bodied insects, and mites. The larvae are caterpillarlike, alligator-shaped, broadest in the thoracic area or anterior part of the abdomen, tapering towards the tail. They are mostly gray, spotted with orange, yellow, or white, and often are found on aphid-infested plants. Many are covered with spines, skin processes, or waxy secretions.

Ladybird beetles are interesting in several ways: when disturbed many species secrete a bitter, amber-colored fluid that is believed to have poisonous effects on vertebrates; some species form huge masses for hibernation (**Pl. 15d**) and migrate long distances to and from hibernation sites and summer feeding grounds; and a number of coccinellids have remarkable color variation in which adults occur in 2 completely different-looking forms (polymorphism). Although rather consistent in body form and habits, there are a great many species, with more than 125 known in California.

388; Plate 15d; cover photo (lower right). Convergent Ladybird. *Hippodamia convergens.* One of the most common insects in California, this species assembles in great numbers (**Plate 15d**) in canyons and hills of the Coast Ranges and Sierra Nevada for hibernation. The beetles are often seen when the overwintering masses begin to disperse on warm days in February or March. They migrate to the valleys where they feed on aphids. Adults of the following generation return to the mountain aggregations in May or June, when the lowland vegetation is drying. ADULT (**cover photo, 388**): BL 6–8 mm; oval, with the wing covers somewhat pointed at the tail end; bright orange with variable spotting; thorax black with a narrow rim and a pair of narrow, diagonal, converging, yellowish spots. LARVA: Elongate, somewhat flattened with the segments strongly protruding; gray with indistinct orange markings on the thorax and larger abdominal spots; feeds on aphids on various plants. RANGE: Throughout all but the driest parts of California, at a wide range of elevations. This species is often confused with the **Five-spotted Ladybird** *(H. quinquesignata).* It is also variable, from an unspotted form west of the Sierra Nevada to one having 3 nearly complete bands across the elytra in the Great Basin. It usually does not have the convergent spots on the thorax. Another similar-appearing, all red species is the **California Ladybird Beetle** *(Coccinella californica).* ADULT: BL 5–7 mm; more convex and less elongate than *Hippodamia* but colored similarly, with pre-

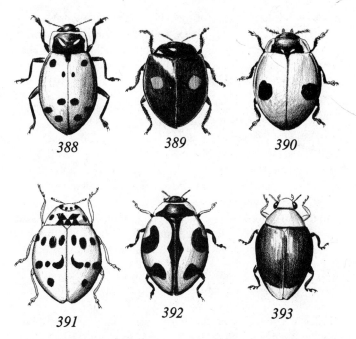

Figs. 388–393 Ladybird Beetles: 388, Convergent Ladybird, *Hippodamia convergens;* 389, Two-Stabbed Ladybeetle, *Chilocorus orbus;* 390, Two-Spotted Ladybeetle, *Adalia bipunctata;* 391, Ashy Gray Ladybird, *Olla abdominalis;* 392, Vedalia, *Rodolia cardinalis;* 393, Mealybug Destroyer, *Cryptolaemus montrouzieri.*

dominately black thorax and reddish orange elytra, usually without spots; color fades to a dull orange or orange-brown in pinned specimens. LARVA: Stout, widest at the second segment, light to dark bluish with orange spots from the fourth abdominal segment forward; body wall with blunt, fleshy protuberances on the dorsum and upper sides; feeds extensively on aphids. RANGE: Coastal counties south to the Transverse Range. Other species of *Coccinella* are similar in appearance, usually with spotting on the elytra, and are more widespread in the state.

389. Two-stabbed Ladybeetle. *Chilocorus orbus.* ADULT: BL 4–5 mm; similar in color to the (**391**) Eyed Ladybird but shiny and more convex, with the head end blunt; commonly black with 2 large, red spots on the elytra, less often pale gray with small, black spots. LARVA: BL to 7 mm; covered with many long, branched spines; black with a median yellow belt; feeds on scale insects of many kinds, especially in orchards and gardens. RANGE: Widespread.

390. Two-spotted Ladybeetle. *Adalia bipunctata*. This small species has distinctly differing color forms. ADULT: BL 4–5 mm; usually red with the thorax black, irregularly margined white, and with a large black spot on each wing cover; other forms are mostly black with 2 variable, orange spots on the front margin and middle of the elytra. LARVA: Slender, widest at abdominal segments 2–4; dark brown to bluish gray, mottled with cream-colored spots; each segment with 3 pairs of raised spots bearing spines; last segment with eversible, fleshy protuberances. RANGE: Widespread; in all but the desert areas.

391. Ashy Gray or **Eyed Ladybird.** *Olla abdominalis*. This species also has 2 distinct forms. ADULT: BL 4–6 mm; nearly hemispherical; either pale tan to grayish with numerous, small, black spots on the thorax and wing covers or a form resembling the Two-Stabbed Ladybeetle, mostly black with 2 large, red spots in the middle of the elytra; distinguished by white on the leading edge of the pronotum. LARVA: BL to 20 mm; black with yellowish markings; feeds on acacia psyllids, aphids of walnut and other plants, etc. RANGE: Widely distributed; primarily at lower elevation and foothill localities.

392. Vedalia. *Rodolia cardinalis*. This is one of the most famous of all insects to entomologists because it was the first successful introduction of a predaceous insect into any country to control a pest insect. This was a breakthrough for applied biological control, which has become increasingly important in the effort to reduce insecticide use. The Vedalia was introduced into southern California from Australia in 1888 and developed huge populations, which within a year effectively controlled the Cottony-cushion Scale, a major pest of citrus. ADULT: Small (BL 2.5-3.5 mm); nearly hemispherical; dark red with irregular black markings which are more extensive in the male; color pattern obscured by fine, gray hairiness.

LARVA: Pink with black markings and a bluish bloom; feeds on Cottony-cushion Scale *(Icerya purchasi)*. RANGE: Established at citrus-growing locations in southern California and the Central Valley.

393. Mealybug Destroyer. *Cryptolaemus montrouzieri.* Following the success of the Vedalia introduction, about 50 more kinds of coccinellids were introduced in the 1890s, but only 4 species, including this one, became established. The Mealybug Destroyer survives the winter only in coastal San Diego County but is mass-produced in insectaries and released by the millions. ADULT: Small (BL: 3–4 mm); oval in outline, tapering to the tail end; shining black with the head, thorax, tips of elytra, and abdomen reddish. LARVA: Yellow, entirely covered with long, white, waxy filaments, resembling large mealybugs. RANGE: Annually released in warmer parts of the state.

LONGHORN BEETLES
Family Cerambycidae

The Cerambycidae includes beetles of a wide array of sizes, colors, shapes, and ornamentation. They have long been a favorite among collectors, probably exceeded in popularity only by butterflies and larger moths. Most cerambycids are somewhat elongate and cylindrical with long antennae, in excess of 2/3 the body length, which are usually attached in a notch of the eye. The beetles have chewing mouthparts, long, rather thin legs, and well-developed wings. Many are rather slow-moving, but some are quick and are seen running rapidly on recently cut wood, or taking flight suddenly. Adults feed on pollen, flower parts, foliage, or bark of twigs, varying according to species; many, especially the brightly colored forms, are found at flowers.

The larvae bore into wood of dying or dead trees or, in some genera, in roots of living shrubs. Different species select moist, dry, or rotting stages of dead wood, while others use bark or wood of living trees. Usually a special pupation chamber is constructed in a gallery leading away from the larval tunnel, and it is plugged by frass and fibrous chips. The larval form is correlated with habits; most are cylindrical, those living in bark are flattened. The head is small and usually somewhat inserted into the prothorax, which is broader than the remaining body segments. The legs are either absent or present, but short. The life history usually requires one year, but in some species involves several years. Larvae of a few species may lie dormant in dry wood for years. There are records of cerambycids emerging 20 years or more later from wood made into furniture or beams in homes.

There are approximately 400 species of longhorn beetles in California, including some of our largest Coleoptera.

394. California Prionus. *Prionus californicus.* ADULT: BL 40–60 mm; similar to the (**397**) Spined Woodborer but broader in form, darker in color, and with sawbladelike, toothed antennae in the male; 3 large spines on each side of the thorax. LARVA: Large, whitish grub somewhat less strongly segmented than the Spined Woodborer and lacking the teeth on the front of the head; lives in dead or dying roots and stumps of oaks and other hardwoods and conifers. RANGE: Coastal and inland valleys, foothills, and mountains to middle elevations.

Plate 15e. Lion Beetle. *Ulochaetes leoninus.* This species has short wing covers and resembles a bumble bee when in flight. ADULT: BL 20–26 mm; black with the thorax densely covered with yellow hair, the short wing covers tipped with yellow, and the legs banded with yellow; hindwings exposed and reach the tip of the upturned abdomen. LARVA: In pines and other conifers. RANGE: Conifer areas of the Coast and Transverse ranges and Sierra Nevada; common in southern California, scarce northward.

395. Oregon Fir Sawyer. *Monochamus oregonensis.* This slow-moving species is not conspicuous but is one of the most spectacular beetles in California. It may be found in numbers on the lower side of recently felled trunks of White Fir and other coniferous trees. ADULT: Large (BL 16–31 mm); enormously elongated antennae, particularly in males; females black with whitish pubescent spots and white annulations on the antennae; males tend to be mostly black. LARVA: In re-

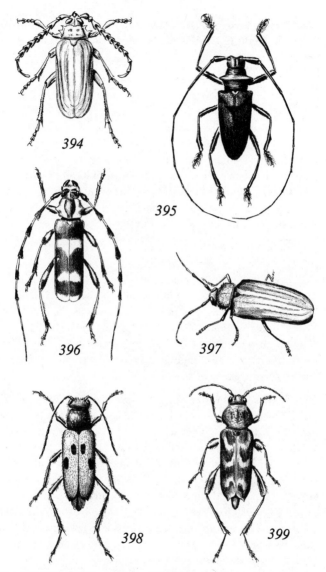

Figs. 394–399 Longhorn Beetles: 394, California Prionus, *Prionus californicus;* 395, Oregon Fir Sawyer, *Monochamus oregonensis;* 396, Banded Alder Borer, *Rosalia funebris;* 397, Spined Woodborer, *Ergates spiculatus;* 398, Dimorphic Flower Longhorn, *Anastrangalia laetifica;* 399, Nautical Borer, *Xylotrechus nauticus.*

cently dead or dying fir, Douglas Fir, and other conifers. RANGE: North Coast Range and Sierra Nevada at middle to higher elevations.

Plate 15f. Red milkweed beetles. *Tetraopes.* The slow-moving, bright red members of this genus are common on milkweed, where they perch conspicuously, evidently protected by distastefulness from vertebrate predators. ADULTS: Blunt, cylindrical with the wing covers broadest at the base, giving a broad-shouldered appearance; red with variable black spotting. LARVAE: In roots of living milkweed *(Asclepias)*. RANGE: *T. basalis* (**Plate 15f**) is the most common species throughout the state west of the Sierra Nevada, while other species inhabit the east side and desert areas.

396. Banded Alder Borer. *Rosalia funebris.* ADULT: Large (BL 25–38 mm); slender with very long antennae (1-2× the BL); beautifully banded, pale bluish white and black. LARVA: In wood of alder, ash, and other hardwoods. RANGE: Coast Ranges. This species was popularly known as the California Laurel Borer, although California Laurel *(Umbellularia)* evidently is not a primary host. The beetles have been observed on several occasions to congregate at freshly painted buildings during hot spells.

397. Spined Woodborer. *Ergates spiculatus.* ADULT: Large (BL 40–45 mm); brown with the head and thorax granulate above and darker than the wing covers; lateral margins of the thorax with many small teeth. LARVA: Large (BL to 60 mm); white grub with strongly differentiated segments and a dark head bearing 4 blunt "teeth"; lives in older, dead logs and stumps of pine and other conifers; maturation requires 2–3 years. RANGE: Coniferous forest regions. This and the (**394**) California Prionus are the largest beetles in California, and they frequently startle residents and campers when they noisily fly into lights on warm summer evenings.

398. Dimorphic Flower Longhorn. *Anastrangalia laetifica.* ADULT: Females are black, with bright red wing covers bearing small black spots; males are either entirely black or have dull yellowish or tan markings on the elytra. The beetles are

very active and visit a wide range of wildflowers in the foothills and mountains during spring and early summer. LARVA:
Unornamented, whitish grub with a brown head; in older, dead
pine wood. RANGE: Mountains and foothills throughout the
state except the desert ranges. Many similarly shaped and
brightly colored species of the same subfamily occur in
California and are common flower visitors. Among these, the
Yellow Velvet Beetle *(Cosmosalia chrysocoma)* is one of the
most conspicuous. It is a beautiful golden yellow, covered with
fine, hairy pile; BL 9–15 mm. Common in mountain meadows
at middle elevations throughout the state.

399. Nautical Borer. *Xylotrechus nauticus.* This beetle is
common on recently felled oak and other hardwoods; it is very
quick and easily startled into flight. ADULT: Medium-sized
(BL 8–15 mm); bullet-shaped with relatively short antennae;
gray or brownish with 3 transverse, irregular white lines on the
elytra. LARVA: BL to 18 mm; white, blunt, with a dark head;
in dead or injured oak, madrone, eucalyptus, and fruit trees.
RANGE: Foothills to middle elevations throughout the state
excepting deserts. A related species, the **Firewood Borer**
(Neoclytus conjunctus), is similar in appearance but has thin,
yellow circular bands. It often emerges from oak firewood in
homes.

LEAF BEETLES
Family Chrysomelidae

In species numbers, the family Chrysomelidae is one of the
largest families of Coleoptera, probably second only to the
Curculionidae, and it is estimated that more than 400 species
are recorded in California. There are various body forms, but
for the most part chrysomelids are small, oval, blunt, and
cylindrical or flattened, and often they are metallic colored.
The antennae are threadlike without a club and are generally
shorter than those of the longhorn beetles, which many leaf
beetles most resemble. Some chrysomelids have enlarged
hindlegs and can jump long distances when startled ("flea
beetles"), but most tend to draw up their legs, feigning death,
and drop when disturbed.

Both the adults and larvae feed on living plants, most often on the leaves. Some are leaf miners in the larval stage and a few bore into stems. Most are fairly specific in host-plant preference. The more commonly seen larvae feed externally on leaves and are caterpillarlike. They are rather fat, tapered towards the tail end, and often metallic colored. During feeding the leaf surface is skeletonized or holes are produced, accompanied by characteristic droppings plastered to the leaf surface. A few species feed on aquatic plants, and still others construct snailshell-like cases of earth and excrement and live in ants' nests or in leaf litter.

400. Western Spotted Cucumber Beetle. *Diabrotica undecimpunctata.* This is one of the commonest insects in California, occurring on a wide array of weedy and cultivated plants. ADULT: BL 4–6 mm; oval in outline with the thorax narrower than wing covers and the head strongly protruding; bright green with 6 variable, black spots on each wing cover and variable amounts of black on legs and underside. They eat leaves and flowers of all kinds except conifers; particularly abundant on plants of the squash family. LARVA: Slender, wiresormlike in form, with short legs; yellowish white with brown head; last segment brown, with a nearly circular outline; in roots of grasses, corn, and other plants. RANGE: Throughout most of California except at the highest elevations.

401. Aquatic leaf beetles. *Donacia.* Beetles of this genus are found on emergent parts of aquatic plants or under water. They breathe by cutting the stem of the plant with their mandibles, allowing oxygen to escape. The oxygen is caught by hairs on the antennae and face and spread back over the body surface. ADULTS: BL 5–9 mm; slender, flattened, with thorax narrower than wing covers; iridescent blue, green, purplish, or bronze colored. LARVAE: Elongate, subcylindrical; whitish; with short, hooked legs. The abdomen is terminated by a pair of spines, enabling larvae to perforate the plant and insert the tail into air spaces for breathing. RANGE: Ponds and lakes at a wide range of elevations.

402. Green Dock Beetle. *Gastroidea cyanea.* ADULT: Small (BL 4–5 mm); oval, convex, bright metallic green; sometimes

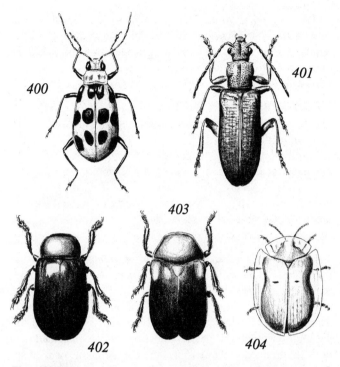

Figs. 400–404 Leaf Beetles: 400, Western Spotted Cucumber Beetle, *Diabrotica undecimpunctata;* 401, Aquatic Leaf Beetle, *Donacia* species; 402, Green Dock Beetle, *Gastroidea cyanea;* 403, Red-Shouldered Leaf Beetle, *Saxinus saucia;* 404, Golden Tortoise Beetle, *Metriona bicolor.*

turning blue in pinned specimens. Common in spring on Curly Dock *(Rumex crispus)* and related plants. The females' abdomens are grossly distended, greatly exceeding the wing covers, before laying large masses of eggs. LARVA: Black, humpbacked grub that eats all but the larger veins of dock leaves. RANGE: Throughout lower elevation parts of the state, west of the mountains.

403. Red-shouldered Leaf Beetle. *Saxinus saucia.* ADULT: BL 4–6 mm; robust, cylindrical; metallic blue above with a conspicuous red spot at the base of each wing cover, pale grayish beneath. Common on buckwheat *(Eriogonum),* ceanothus, and other flowers, where they eat the buds.

LARVA: Slender, white; in cases made of mud and frass in ants' nests. The eggs are brown with longitudinal ribs, resembling seeds, and are probably collected by ants. RANGE: Throughout foothill and lower elevation parts of the state, excluding the deserts; occasionally at higher elevations.

Plate 15g. Blue Milkweed Beetle. *Chrysochus cobaltinus.* This is the largest and most spectacular chrysomelid in California. ADULT: BL 8–12 mm; robust, oval, with the thorax slightly narrower than the wing covers; beautiful, shining, metallic dark blue to greenish blue. Common on milkweed *(Asclepias).* LARVA: White grubs in the soil, feeding on roots of the milkweed (presumed from habits of related eastern species). RANGE: Throughout the state at a wide range of elevations, wherever milkweed is found.

404. Golden Tortoise Beetle. *Metriona bicolor.* These and related beetles look like burnished brass buttons and are sometimes called "gold bugs." ADULT: BL 5–6 mm; flattened, nearly round in outline with the thorax and wing covers expanded so as to completely conceal the head and legs. In life the shell is golden and nearly translucent in the expanded portions, but pinned specimens become dull ochreous or reddish. The color can be restored temporarily by placing the specimens in a humidity chamber. LARVA: Mealybuglike; oval; brownish with branched spines on the margins; carry fecal material on the back, which is supported and attached by a forked tail process. Larvae and adults eat the leaves of morning glory and related plants, but the adults are found on various plants. RANGE: Foothills and valleys west of the mountains.

WEEVILS
Family Curculionidae

No other family in the entire animal kingdom has as many species as the family Curculionidae. The numbers and diversity of form and habits of weevils are legion; probably there are more than 1,000 species in California. The major distinguishing features are the long snout, which may be slender or broad, and the elbowed, clubbed antennae, which arise from the sides of the snout. The body shape and ornamentations are exceedingly varied—oval, elongate, flattened, with long or short legs, and varying in size from tiny to large (1 to 35 mm in length). In general, weevils use the snout in boring out a pit for the eggs. Unlike most Coleoptera, many weevils are clothed with scales. Most North American forms are dull colored, but in tropical regions many are resplendent in bright, metallic colors.

The larvae are legless grubs and generally similar in form. The majority feed inside seeds, stems, or roots, while some subterranean species eat the surface of roots, and a few live on aquatic plants. Usually pupation occurs in the soil or in larval galleries, although some genera produce a parchmentlike cocoon. Many curculionids cause serious economic damage to crops or stored foods.

405. California Acorn Weevil. *Curculio uniformis.* ADULT: BL 5–7 mm; robust, with a narrow thorax, long legs, and greatly elongated snout, in the female much longer than the body, shorter in the male; body brown-spotted, covered with bright yellow and ochreous scales. LARVA: In acorns of several species of oaks. RANGE: Common on live oaks and deciduous oaks in the foothills, especially during the bloom and early acorn periods.

406. Rose Curculio. *Rhynchites bicolor.* ADULT: Small (BL 4–6 mm); rather slender with the thorax elongated, much narrower than the wing covers; body black with the elytra and back of the thorax bright red. The beetles bore holes for feeding and egg-laying in the buds of plants of the rose family. LARVA: Small, white grubs in the base of flowers and seed pods; in roses, blackberry, and related plants. RANGE: Throughout foothill to middle-elevation areas of the state.

407. Granary Weevil. *Sitophilus granarius.* ADULT: Small, slender (BL 3–4 mm); beak and thorax elongate, each nearly as long as the wing covers; moderately shining, pale to dark brown; thorax with numerous small depressions; wing covers with several parallel linear ridges; flightless. LARVA: Stout, ovoid, white grub with a dark head; in kernels of grain or caked pieces of grain products. RANGE: Urban and agricultural

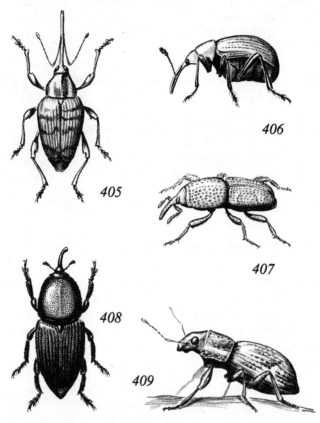

Figs. 405–409 Weevils: 405, California Acorn Weevil, *Curculio uniformis;* 406, Rose Curculio, *Rhynchites bicolor;* 407, Granary Weevil, *Sitophilus granarius;* 408, Yucca Weevil, *Scyphophorus yuccae;* 409, Fuller's Rose Weevil, *Pantomorus godmani.*

areas. This beetle has been carried by commerce to all inhabited parts of the world and probably is not native to North America. A similar species, the **Rice Weevil** *(S. oryzae),* is less common in California, but occurs widely in warmer areas, infesting various seeds and grains. It is a little smaller and has relatively shorter wing covers, which usually have 4 pale areas. The life history and habits are similar to the Granary Weevil, but adults of the Rice Weevil are able to fly.

408. Yucca Weevil. *Scyphophorus yuccae.* One of the largest weevils in California, this species is abundant at the bases of Quixote Plant *(Yucca whipplei)* and on Joshua Tree *(Y. brevifolia).* ADULT: BL 12–20 mm; black; rectangulate in outline, somewhat flattened; with thickened legs and beak; thorax smooth, as broad as the wing covers which are deeply etched with longitudinal grooves. LARVA: In green, flowering stalks of yucca, leaving before the stalks dry to pupate in the ground. RANGE: Throughout the range of yucca in California. A related species, the **Tule Billbug** *(Sphenophorus discolor),* is similar in shape and size (BL 14–18 mm) but is somewhat rounded in outline and has a shiny surface. It is black above with variable whitish streaks and mostly whitish or pale tan below. It normally breeds in tules or cattails along rivers, but adults sometimes cause damage in grain fields nearby. RANGE: Along permanent rivers and creeks of the Central Valley and coastal areas.

409. Fuller's Rose Weevil. *Pantomorus godmani.* ADULT: BL 7–9 mm; cylindrical, broad-nosed; dull brown, faintly covered with a whitish, waxy powder and marked with a white, oblique stripe on each side of the sealed elytra; flightless. The adults hibernate and often appear in large numbers in spring, eating the leaves of a wide variety of plants; frequently seen on sidewalks and driveways in cities in fall and spring. LARVA: BL 5–6 mm; white; legless grubs; in the roots and lower stems of various plants, often a pest of ornamental garden plants. RANGE: At lower elevations throughout the state, excepting the deserts.

BARK BEETLES
Family Scolytidae

Scolytids are small, dull-colored, cylindrical, short-legged beetles. They live in the bark or between the bark and wood of trees and shrubs. Their secretive habits and inconspicuous appearance belie their importance, for bark beetles kill more trees than all other forces combined, including forest fires.

The adults bore an entrance tunnel into bark or, in a few species, into seeds or cones. At the inner end of the tunnel, 2 or more egg tunnels are cut between the bark and the wood, and eggs are scattered or laid in niches along the walls. Later the white, legless larvae excavate slender burrows at right angles to the egg tunnels. These galleries become filled with frass and increase in diameter as the larvae grow. When population densities are high, the galleries eventually encircle the limb or trunk and cut off flow of water and food in the critical cambium layer, killing the parts of the trees beyond that point. The form or arrangement of the galleries (**Plate 15h**) is characteristic for given species or genera, and practiced forest entomologists can identify specific bark beetles by them.

There is remarkable variation in relations of the sexes, with some species polygamous, others monogamous. In some, a nuptial chamber is excavated by the male, and he produces a powerful scent in the frass, which attracts females of the same species.

Some scolytids, called ambrosia beetles, feed deeper in the wood. Females prepare a bed or layer of chips and frass, upon which fungus develops to be used as food by the beetles and larvae. Ambrosia—food of the gods—is the name applied to this fungus-food.

California, owing to its rich diversity of plant communities, has a large fauna of bark beetles, with more than 170 species in 44 genera.

410. Red Turpentine Beetle. *Dendroctonus valens.* ADULT: Moderately large (BL 6–9 mm), robust; bright reddish brown. Galleries occur near the base of all kinds of pines. RANGE: Sierra Nevada and Coast Ranges; into low-elevation urban areas, where it sometimes kills ornamental trees, such as Monterey Pine where it is grown in hotter, drier zones than its native range along the coast. A related species, the **Western Pine Beetle** *(Dendroctonus brevicomis),* known as "DB" to forest entomologists, is the most destructive insect to pines in the western United States. ADULT: Small (BL 3–5 mm); broadly cylindrical, pale brown to black. The female cuts long, meandering egg galleries with egg niches 8–10 mm apart in Ponderosa and Coulter pines. RANGE: Throughout the Sierra Nevada and scattered in the Coast Ranges.

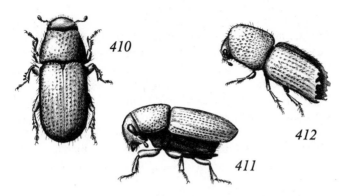

Figs. 410–412 Bark Beetles: 410, Red Turpentine Beetle, *Dendroctonus valens;* 411, Smaller European Elm Bark Beetle, *Scolytus multistriatus;* 412, California Five-Spined Ips, *Ips paraconfusus.*

411. Smaller European Elm Bark Beetle. *Scolytus multistriatus.* ADULT: Small (BL 3–4 mm); dark brown or blackish; distinguished by a long spine that arises from the underside of the second abdominal segment of the male. RANGE: Cities and suburban areas of the San Francisco Bay area and southern California. It was accidentally introduced to North America around the turn of the century and first appeared in California in 1951. Its feeding causes little damage, but the beetles transport a fungus which causes Dutch Elm Disease, long a serious killer of trees in the eastern United States. The disease spread to California in 1974 and is subjected to eradication attempts. Another species, the **Shothole Borer** *(S. rugulosus),* was introduced from Europe and now occurs in many parts of North America. This scolytid feeds in pome and stone-fruit trees and frequently causes limbs to die in orchards. ADULT: Small (BL 2–3 mm); exceptionally stout; nearly black with the antennae and legs cinnamon red. RANGE: Coastal areas and valleys and Central Valley.

412. California Five-spined Ips or **Yellow Pine Engraver.** *Ips paraconfusus.* Species of *Ips* have a concave area at the tail end of the elytra, bearing spines on its margins. The genus includes some of the most important insects in the forest industry in California and the most destructive of them is this species. ADULT: Small (BL 4–5 mm); brown; with 5 spines on each lateral margin. The male is joined by 3–5 females, each of which bores an egg gallery along the grain of the wood. The resultant pattern resembles a tuning fork. LARVA: In apparently healthy smaller trees, damaged trees, and tops of larger trees of all species of pine within its range. RANGE: West side of the Sierra Nevada, Coast, and southern ranges. This species was differentiated only recently from **Pinyon Pine Engraver** *(I. confusus),* and nearly all older literature on *I. confusus* refers to *I. paraconfusus.*

STYLOPS
Order Strepsiptera

The few members of the order Strepsiptera are seldom-seen, parasitic insects. The larvae feed inside wasps, bees, Homoptera, Diptera, Hemiptera, orthopteroid insects, or silverfish. Usually the adult female spends all her life in a case inside the host insect, which is said to be "stylopized." The males are small creatures with thick, featherlike antennae, bulging eyes, small, twisted forewings, and large, fanlike hindwings. Fewer than 30 species are known in California.

Family Stylopidae

413, 414, 415. Pacific Stylops. *Stylops pacifica.* This species parasitizes andrenid bees, and although the males are short-lived and rarely seen, stylopized bees are fairly common, especially early in spring. ADULT: Male (**414**) small (BL 2–3 mm); black with white hindwings. They may be attracted in numbers to a caged, stylopized bee if the female stylops is unmated. Females (**415**) grublike, evidenced only by the tip of the head and thorax protruding from between adjacent abdominal segments of the host bee. LARVA: When newly hatched, rapidly running, minute, silverfish-shaped (triungulin) (**413**). Larvae emerge from the living bee as it visits flowers, where they wait to attach to legs of other bees that come along, to be carried back to the nests. There they seek out the bee larvae, burrow in, and change to a legless maggotlike form. They are fed by the host's blood and go through several molts but do not

413   *414*   *415*

Figs. 413–415 Stylops: 413, Pacific Stylops, *Stylops pacifica* (triungulin); 414, *S. pacifica* (male); 415, *S. pacifica* (female).

kill the bee larva. After the host pupates and undergoes metamorphosis to the adult bee, the stylops protrudes its head between 2 abdominal segments. Adult male stylops emerge through the projecting pupal shell, while the larviform female remains in place. Either sex of the bee may be stylopized and up to 30 stylops of both sexes may occur in a single bee. RANGE: Coast Ranges.

ANTS, WASPS, BEES
Order Hymenoptera

Members of the order Hymenoptera are insects with 2 pairs of membranous wings, often with the venation greatly reduced, or they are wingless in one or both sexes, certain castes, or in certain seasons. The mouthparts are adapted for biting, and the abdomen is usually strongly constricted at the second segment with the basal segment (propodeum) fused to the thorax. Females are equipped with an ovipositor which is developed for sawing or piercing to insert eggs into plants or into other insects, or it may be modified for stinging. Ants, wasps, and bees inject a mixture of proteins and enzymes which act on the tissues of the victim as toxins or cause the release of histamine. In humans, depending upon individual sensitivity, the effects may be severe. Most painful ''bites'' caused by insects actually are the stings of female Hymenoptera. Few members of this order bite strongly enough to affect humans, and the males do not sting.

The Hymenoptera is an enormous order, with the second largest number of described species, and many thousands of forms awaiting discovery. Many Hymenoptera live during the larval stage as parasitic forms (parasitoids) within the bodies of other insects. There are many families of these parasitoids with great diversity, varying in structure and habits according to the family and kinds of insects parasitized. Many are tiny, little-known insects, and it is likely that more than half the species of this order will ultimately be described as parasitoids. In nonparasitic forms there is a great diversity of behavioral adaptation and structural modification: sawflies produce caterpillarlike larvae which feed on leaves; wasps paralyze insects or spiders and transport them to nests to feed their legless, grublike larvae; bees gather pollen for their young. In many ways, however, specializations and life styles have reached their highest development in the socially organized groups: ants, certain wasps of the family Vespidae, and bees of the family Apidae. A large proportion of these societies constitute a separate type or caste known as the worker. These are modified females which do not lay eggs. Their functions include nest-building, food-gathering, tending of the nurseries, and so on. Defense of the colony may be another task of workers and, in ants, may be relegated to a special ''soldier'' caste. There is a great deal of variation in complexity and habits among semi-social and social forms.

Among solitary wasps and bees, females may provision the nest progressively, continuing to feed larvae from time to time; more often, mass provisioning is practiced, with each cell completely stocked and closed off before the egg hatches.

There is no accurate estimate of the number of Hymenoptera species in California, but the described species alone probably exceed 5,000, with most of the small, parasitic forms not yet named.

References

Bohart, R. M., and R. C. Bechtel. 1957. The social wasps of California. *Calif. Insect Survey, Bull.* 4(3):73–102.

Creighton, W. 1950. The Ants of North America. *Mus. Comparative Zool., Harvard, Bull.* 104; 585 pp.

Duncan, C. 1939. A contribution to the biology of North American vespine wasps. *Stanford U. Publ. Biol. Sci.* 8; 272 pp.

Evans, H. E. 1963. *Wasp Farm.* New York: Natural History Press.

Evans, H. E., and M. J. Eberhard. 1970. *The Wasps.* University of Michigan Press; Ann Arbor; 265 pp.

Michener, C. D. 1974. *The Social Behavior of Bees. A Comparative Study.* Cambridge, Mass.: Belknap, Harvard University Press; 404 pp.

STEM SAWFLIES
Family Cephidae

Cephids are small, slender-bodied insects with a large thorax and the abdomen flattened side-to-side. They are mostly yellow and black, with transparent wings. The larvae bore in the stems and shoots of various plants and are legless except for tuberclelike rudiments. The abdomen ends in a small, retractile spine. Adults are common at flowers in spring; only 5 species are known in California.

416. Wheat Stem Sawfly. *Cephus cinctus.* ADULT: BL 8–11 mm; black with variable yellow markings and transparent wings. LARVA: BL 12 mm; thorax somewhat enlarged; pale yellow with a brown head; in stems of various grasses, near the ground. RANGE: Widely distributed in coastal and mountain areas, absent from the San Joaquin Valley and deserts. A similar species, the **Western Grass Stem Sawfly** (*C. clavatus*), occurs in the same range. The two occasionally are found together; *C. cinctus* emerges earlier in the spring and has a longer flight period.

SAWFLIES
Family Tenthredinidae

Sawflies are small to medium-sized, somewhat flattened insects with a rather wide abdomen, broadly attached to the thorax. The ovipositor is adapted for sawing, and females usually cut a slit in young shoots or leaves and insert the eggs. The larvae of most species feed externally on leaves and resemble caterpillars of Lepidoptera, from which they are distinguished by having false legs on all segments of the abdomen (these are always lacking from at least segments 1–2 in Lepidoptera) and

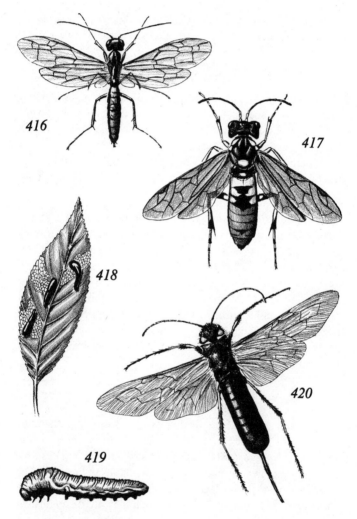

Figs. 416–420 Sawflies: 416, Wheat Stem Sawfly, *Cephus cinctus;* 417, Common Sawfly, *Tenthredo* species; 418, Pearslug, *Caliroa cerasi* (older larvae on leaf); 419, *C. cerasi* (newly molted larva, lateral view); 420, California Horntail, *Urocerus californicus.*

by having a single, large "eye" on each side instead of the group of small eyespots found in Lepidoptera. Some sawfly larvae cause plant galls in which they feed.

Female tenthredinids often produce eggs without mating, in which case all or most resulting offspring are females. Males are unknown in many species. Although rather uniform in structure, the family is large and its members feed on a wide diversity of flowering plants and ferns. There are more than 100 species known in California.

417. Common sawflies. *Tenthredo*. There are many species of this genus in California. ADULTS: Rather large (BL 8–14 mm); entirely black or rust-red or black with yellowish green transverse bands and a spot on the hind part of the thorax, and/or the distal half of the abdomen rust-red; head large and quadrate, abdomen flattened and broadened towards the tip. RANGE: Throughout all but the most arid parts of California. LARVAE: Caterpillarlike; wide variety of colors and markings; feed on diverse plants according to the sawfly species; often curl and drop, feigning death, when disturbed. The **Two-Gemmed Sawfly** *(T. bigemina)* is common in the Sierra Nevada at middle elevations in spring. It is black with bright yellow or yellow-green bands. **Green willow sawflies** of the genus *Rhogogaster* are similar to *Tenthredo* in size and form. There are 3 common species in California. ADULTS: Medium-sized (BL 7–11 mm); bright green with a broad band of black markings down the back, and transparent wings. LARVAE: Slender, caterpillarlike, of various colors. They live solitary on willow and other broadleafed plants. RANGE: Lower to middle elevation areas of the state outside the deserts.

418, 419. Pearslug. *Caliroa cerasi*. ADULT: BL 6–8 mm; shining black with darkened wings. LARVA (**418, 419**): BL 10–13 mm; yellow when freshly molted, becoming dark olive green owing to a covering of slimelike material, which causes it to look like a slug. Feeding causes skeletonizing of the leaf surface; on cherries, pear, plum, etc. RANGE: Mountains and fruit-growing districts of the state. Pear orchards also are affected by a similar species, the **California Pear Sawfly** *(Pris-*

tiphora abbreviata). ADULT: BL 5–6 mm; black with yellow markings on the thorax, and transparent wings. LARVA: BL 13 mm; bright green, matching the leaves on which they feed; on pear trees in spring. They have a characteristic habit of resting on the margins of leaves in oval holes eaten into the edges. RANGE: Northern half of California.

Plate 16a. Willow Leafgall Sawfly. *Euura pacifica.* The red, berrylike galls caused by this and related species are a familiar sight on willow leaves (**Plate 16a**). ADULT: BL 3–5 mm; body shining black in male, dull reddish in the female; wings transparent. LARVA: Whitish, grublike; feeding entirely within the galls, leaving to form tough, silken cocoons on the ground. RANGE: Widespread at low to middle elevations.

PARASITIC WOODWASPS
Family Orussidae

Plate 16b. Western Orussus. *Orussus occidentalis.* This insect has unique habits. Adults patrol on bark-free wood of fallen logs of varying ages, where they resemble carpenter ants. Females detect galleries of wood-boring larvae, possibly by differences in reflected vibrations, when they drum their antennae on the surface. Oviposition occurs by drilling through 5–20 mm of sound wood. After the eggs hatch, the larvae travel through the frass-packed gallery to the host occupant, where they feed on it, beginning at its tail end. ADULT: BL 8–14 mm; stout wasp with short wings; bluish black with the last 6 segments shining reddish. LARVA: White; legless, somewhat flattened grub with the tail end upturned but without a spine; feeding externally on larvae or pupae of horntails, flat-headed borers, or longhorn beetles. RANGE: Coniferous forest areas of the Coast Ranges and Sierra Nevada.

HORNTAILS
Family Siricidae

Horntails are large, brightly colored insects. The body is cylindrical with the head, thorax, and abdomen of about the same width. The abdomen ends in a spine or horn which is short and triangular in males, lanceolate in females. The ovipositor is strong and, when at rest, projects backward with the plane of the body, resembling a huge stinger. However, horntails do not sting; the ovipositor is used for drilling into wood, where the eggs are deposited. The larvae have large heads, reduced legs, and a hornlike process on the last segment. They feed in the heartwood of scorched, weak, or recently killed trees. These insects are carried in shipments of timber or cut lumber, and the adults sometimes emerge after several years, causing holes in walls or flooring in homes. About a dozen species occur in California.

420. California Horntail. *Urocerus californicus*. This is the largest and most commonly collected western siricid. ADULT: BL including horn, 18–40 mm in female, 12–25 mm in male; wings, antennae, and leg bands yellow-orange; body in females entirely blue-black except for a pale spot behind each eye, in males entirely rust brown, the thorax darker. LARVA: Yellowish white; slightly S-shaped with a short, sharply pointed spur on the tail end; in wood of fir *(Abies)* and occasionally other conifers. RANGE: Coniferous forest areas of the state.

BRACONID WASPS
Family Braconidae

Braconid wasps are small, stout or slender, with long antennae and transparent wings. In general structure and habits they are similar to ichneumonids, but lack the second, recurved vein present on the outer part of the forewing in ichneumonids. Generally, braconids are smaller wasps with a larger dark spot or stigma on the leading margin of the forewing. Many braconids are relatively stout-bodied, with antennae shorter than the forewing.

All braconids live as parasitoids in the larval stage, and a great variety of insects serve as hosts. Lepidoptera larvae are the most commonly parasitized, and numerous individual braconids may emerge from a single caterpillar. The larvae are cigar-shaped, with a large head in the first instar. Forms that are internal parasites often have a tail appendage. When larvae are mature, they usually gnaw their way through the host's body wall and construct an external silken cocoon. Cocoons are usually white or yellowish and somewhat fluffy in appearance, rather than smooth and paperlike as in Ichneumonidae.

This is an extremely diverse family, with considerable habitat specialization among its members. Some parasitize Diptera in particular habitats, others use larvae of wood-boring Coleoptera, weevil grubs, adult beetles, leaf miners, or aphids. Several hundred species are known in California, and probably even greater numbers are not yet described.

421. Common braconids. *Apanteles.* Members of this genus parasitize a wide variety of lepidopterous larvae, including leaf miners, small leaf-rollers, cutworms, etc. In larger hosts, such as sphinx moth caterpillars, *Apanteles* sometimes are gregarious, and 50 or 100 white cocoons appear on the back and sides of the dying victim. ADULTS: Small (BL 2–3 mm); black, sometimes marked with reddish; with a short abdomen relative to the wings and legs. LARVAE: Whitish with reduced head and swollen anal segment but no tail-like extension; live inside Lepidoptera larvae. COCOONS: External, silken, cylindrical, white or yellow, sometimes matted together in a heap. RANGE: Virtually throughout the state.

422. Aphid parasites. *Aphidius.* All species of this and related genera parasitize aphids, one wasp per host. When about to pupate, the larva makes a cut in the bottom of the aphid and cements it to the leaf. Such carcasses are familiar objects, pale colored, each with a large circular hole through which the wasp has emerged. ADULTS: Tiny (BL 1–2 mm); slender with reduced wing venation; black with brownish or yellow markings on legs and abdomen, or entirely amber colored. Females have a long, thin abdomen which is brought forward, between the legs, when depositing the eggs into an aphid in front of the wasp. LARVAE: Inside aphids, forming the cocoon there. RANGE: Throughout most of the state.

ICHNEUMONIDS
Family Ichneumonidae

In terms of described species, the Ichneumonidae is the largest family of Hymenoptera. If all the world's species were known, it might well be the largest family of animals. Probably

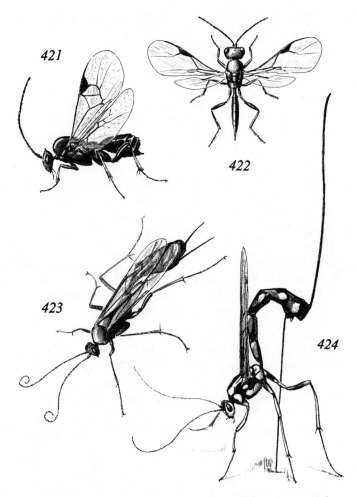

Figs. 421–424 Parasitic Wasps: 421, Common Braconid, *Apanteles* species; 422, Aphid Parasite, *Aphidius* species; 423, Common Ichneumonid, *Ophion* species; 424, Stumpstabber, *Megarhyssa nortoni*.

more than 1,000 species occur in California. All are parasites of insects or occasionally other arthropods. Therefore, their role in natural and man-affected communities is of enormous importance, and the majority are beneficial to man's economy. Most ichneumonids are day-fliers, and general collecting yields a variety of species.

Adult ichneumonids are small to large, slender insects with a threadlike waist, exceptionally long antennae, and transparent wings. They inhabit every ecological situation and prey upon all kinds of insects, specializing according to genera and species. The majority are parasites of Lepidoptera larvae, but many use either wood-boring or other Coleoptera, Diptera, wood wasps, aphids, or spiders, and some are parasites of other ichneumonids or other parasitic insects (hyperparasites). Usually the female searches a particular habitat, such as leafrolls or similar caterpillar shelters on trees, and oviposits into the body of any suitably sized victim she find there. Thus many ichneumonids are not specialists on particular genera or families of host insects, but parasitize a diversity that occupy similar habitats. A single larva develops within each host. When newly hatched, the larva has a prominent, tail-like extension. The full-grown larvae are tailless, variably shaped, legless grubs with a greatly reduced head. Pupation usually occurs outside the host's body in a smooth, oval cocoon, but often it takes place within the host body, especially in species whose larvae delay maturing until the host has reached its pupal stage.

423. Common ichneumonids. *Ophion.* The familiar orange wasps of this genus and several closely related genera are often attracted to lights, even during winter when little else is flying. ADULTS: Moderately large (BL 10–28 mm); body amber colored to reddish brown, sometimes with yellow markings; very slender, with the abdomen flattened side-to-side; wings transparent; legs and antennae long (antennae often 1.25 FW length). Although frightening in appearance, they do not sting. LARVAE: Live inside cutworms and other large caterpillars and scarab beetle grubs. RANGE: Throughout most of the state at low to moderate elevations, especially in arid areas. Members of a genus which is not closely related, *Netelia,* are similar in general appearance and habits, being nocturnal parasites of cutworms. They are often attracted to lights and are sometimes confused with *Ophion,* but *Netelia* can deliver a formidable sting. They are usually brick reddish and have shorter legs and antennae (shorter than FW).

424. Stumpstabber. *Megarhyssa nortoni.* This is the largest ichneumonid in California. Known by a less delicate name to woodsmen, the female wasps are seen drilling their long ovipositors into stumps and recently felled logs, where they deposit eggs in galleries of siricid wood wasps. ADULT: Large (BL 25–38 mm, with ovipositor another 50–75 mm); marked with black, red-brown, and yellow, the legs mostly yellow. LARVA: Feeding externally on nearly full-grown larvae of wood wasps. RANGE: Middle to higher elevations of coniferous forest areas.

GALL WASPS
Family Cynipidae

Most members of the family Cynipidae cause plant galls, which provide shelter and food for the larvae, or they live as guests (inquilines) in galls of other cynipids. The larvae consequently are internal feeders, maggotlike, without legs or a well-defined head. Pupation occurs in the larval cell without a cocoon. Adult cynipids are stout, with the abdomen oval, somewhat compressed and shining, its segments telescoped. The body surface frequently appears smooth and polished, and the wings have few veins.

Galls caused by these insects are exceedingly varied in form and location on plants, but the gall of any given cynipid is of characteristic shape. Gall formation is not well understood. In some cases the gall does not begin to grow until weeks or months after the wasp deposits the egg, so probably a secretion of the newly hatched larva influences abnormal growth of cells in the plant. In many cynipids there is a marked difference between seasonal generations of the wasp, both in body form and in the characteristic gall. One generation produces only females, which oviposit in a different part of the host and give rise to a bisexual generation.

The Cynipidae includes a large number of species, but the host plants are surprisingly restricted, with about 85% of described species on oaks and another 7% on roses. A small proportion of cynipids are parasitic on dipterous puparia or hymenopterous aphid parasites. Upwards of 200 species are recorded in California, and many types of galls are known which have not yet been associated with their wasps.

425. California Oak Gall Wasp. *Andricus californicus.* This species causes the largest and best known gall in the western United States. It begins to form on twigs of Oregon and Valley oaks in spring and is at first kidney-shaped and green, becoming apple-shaped and up to 10 cm long, pale green often with a reddish side ("oak apples"). By late summer the gall dries and turns tan. ADULT: BL 3–5 mm; reddish brown with transparent wings. LARVAE: Few to many in each gall, in individual cells near the center. They mature in fall and the wasps deposit eggs which overwinter. Collections of galls yield several kinds of lepidopterous and coleopterous inquilines as well as several

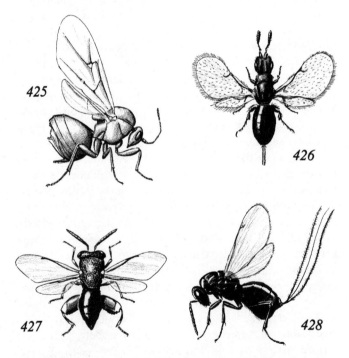

Figs. 425–428 Wasps: 425, California Oak Gall Wasp, *Andricus californicus;* 426, Fig Wasp, *Blastophaga psenes* (female); 427, Butterfly Chrysalis Chalcid, *Brachymeria ovata;* 428, Oak Gall Chalcid, *Torymus californicus.*

parasitic Hymenoptera. RANGE: Coast Ranges, foothill and valley areas.

CHALCIDS
Superfamily Chalcidoidea

The chalcids are a diverse assemblage of small to minute wasps, made up of more than a dozen families, each with interesting, specialized structural and behavioral modifications. Chalcids are numerous in kind, with several hundred described species in California and probably still greater numbers unnamed. They often occur in great individual numbers, with many capable of producing a large number of offspring from a single egg (polyembryonic). Yet chalcids are mostly overlooked because of their small size.

Most species are parasitic in other insects, including other parasites (hyperparasitic). They are of even greater importance than ichneumonids and braconids in natural control of insect populations owing to their extremely specific host preference and high fecundity.

A few groups are plant-feeding and the larvae live in seeds or in plant galls which they cause. The parasitic forms most commonly affect Lepidoptera, Homoptera, Hemiptera, Diptera, or other parasitic Hymenoptera. Among the Chalcidoidea are some of the smallest of all known insects, including parasites of other insects' eggs, less than one-half of 1 mm in length in the adult stage. Adults of most species are stout-bodied, often metallic colored, and have very few veins in the wings.

Family Agaontidae

426. Fig Wasp. *Blastophaga psenes*. This is one of the most remarkable of all insects in structure and biology. The sexes are totally dissimilar and highly specialized for life in fig flowers. The eggs are laid in the ovaries of the caprifig variety (which contains both male and female flowers) and galls are produced. The wingless male wasp emerges first, and upon locating a gall containing a female, gnaws a hole in the gall and fertilizes her. The female then leaves the flower and, laden with pollen, flies to a nearby fruit. The Smyrna variety of fig has no male flowers, and its pollination is affected by "caprification," a process of hanging caprifigs in Smyrna trees. Should the female wasp enter a Smyrna fig, she finds the shape of the flowers prevents oviposition, but wandering about, she incidentally pollinates the flowers. ADULT: BL 2–2.5 mm; females (**426**) are shining black, winged; males are amber colored, uniquely modified, wingless, larvalike with an elongate abdomen that is normally curled under the body, extremely short antennae, greatly enlarged basal leg segments. LARVA: Tiny, legless grubs; white with brown mandibles. RANGE: On cultivated figs.

Family Chalcididae

427. Butterfly Chrysalis Chalcid. *Brachymeria ovata.* This rather large and colorful species often emerges from pupae of butterflies or moths, to the chagrin of the expectant lepidopterist. ADULT: BL 3–7 mm; robust; black and yellow; with a greatly enlarged hindleg segment that is black with a white or yellow spot at its tip. RANGE: Widespread at low to middle elevations, parasitizing a wide variety of caterpillars.

Family Torymidae

428. Oak Gall Chalcid. *Torymus californicus.* This spectacular little wasp parasitizes larvae of cynipid wasps that cause galls on oaks. ADULT: Small (BL 2–3 mm), stout; iridescent green, reflecting gold and coppery in varying lights, with black antennae and reddish legs; females have a long, thin ovipositor extending 4–6 mm beyond the body. RANGE: Foothill oak woodlands, associated with Blue, Valley, and Live oaks.

CUCKOO WASPS
Family Chrysididae

Chrysidids are metallic green, blue, or reddish wasps with a hard, often coarsely sculptured integument. They are easily recognizable by the abdomen which is a little longer than the head and thorax combined; the abdomen is convex above and slightly concave, smooth, and shining beneath. When the wasp is alarmed it brings the abdomen forward, closely appressing it to the thorax, so that the insect rolls itself into a ball like a sowbug does. Cuckoo wasps fly during sunny, warmer parts of the day. Their common name refers to the analogy with cuckoo birds: their larvae are raised in the nests of other wasps and bees. When a host burrow is left unguarded, the female chrysidid enters to deposit an egg. The resulting larva is either an external predator on the larval occupant of the nest, sometimes after the host cocoon is spun; or the newly hatched chrysidid kills the rightful occupant of the cell in an early stage and feeds on the provisions supplied by the unsuspecting nest-builder. The larvae are fat, legless grubs, superficially indistinguishable from those of their wasp and bee hosts.

Cuckoo wasps tend to be specialized in habitat preference, such as ground-nesting vs. hollow-stem-nesting hosts, but are indiscriminate as to the kind of host encountered there. Each chrysidid feeds in a single host cell, and as a consequence, there is considerable size variation within a species, depending upon the size of the host. Those described here are among the largest of more than 50 species known in California.

Plate 16c. Pacific Cuckoo Wasp. *Chrysis pacifica.* This is one of the most widespread and common chrysidids in California. ADULT: Variable in size (BL 4–9 mm); metallic green to deep blue; integument densely punctured, with a row of deeper dents, like teeth marks, preceding the rim of the last visible abdominal segment. LARVA: Attaches to larva of a (**450**) metallic leafcutter bee *(Osmia)* and feeds after the host cocoon is spun. RANGE: Throughout mountains of the state, from foothills to timberline. A related species, the **Large Blue Cuckoo Wasp** *(Chrysis coerulans),* preys on the nests of solitary vespid wasps that use holes in twigs or wood or abandoned mud-dauber nests. The egg is deposited during host provisioning; the *Chrysis* larva destroys the egg of the vespid, then feeds on the paralyzed caterpillar provisions. ADULT: Small to moderately large (BL 6–11 mm); dark green to blue with densely punctate integument and a 4-toothed, flared extension on the abdominal tip. RANGE: Widespread in coastal, valley, and mountainous parts of the state. Another large, common species is **Edwards' Cuckoo Wasp** *(Parnopes edwardsi).* ADULT: BL 6–12 mm; brilliant green to deep blue with densely punctured integument, pronounced depressions between the abdominal segments, and 2 deep creases on the last segment. LARVA: In burrows of solitary ground-nesting wasps. RANGE: Widespread in diverse habitats, from the Central Valley to the high Sierra.

Family Tiphiidae

Tiphiids are typically slender wasps, the females often wingless. They are not often seen by the general public, although some are very common in desert areas. All species are external parasites in the larval stage, usually on scarab or tiger beetle larvae in the soil. Winged adults commonly feed at flowers, on aphid honeydew, or on other exudates on plants. Whether winged or not, the female wasps enter the soil to seek out prey on which to oviposit. About 50 species are recorded in California, but little or nothing is known about the biology of most of them.

429. Tiger tiphiids. *Myzinum.* These are the largest tiphiids in California. ADULTS: BL 10–18 mm; males are slender with elongate antennae, wings, and abdomen; black and yellow striped; females are larger, much broader, and have broadened legs adapted for digging. LARVAE: Probably prey on scarabaeid larvae in the soil, as is true of related species in the eastern United States. Larval growth takes only a few days. RANGE: Desert areas and north in the foothills and Central Valley at least to Antioch, where the 2 common species, *M. frontale* and *M. maculatum,* occurred before 1955 but are nearly extinct now owing to destruction of the sand dunes.

430. Nocturnal tiphiids. *Brachycistis.* This and related genera are represented by numerous species in California. At times males are attracted to lights in large numbers, yet their biology remains one of the unsolved mysteries in insect natural history. The females are wingless and so different in form that most were originally described in separate genera from the males. ADULTS: Males resemble flying ants, reddish brown to black, shiny, slender wasps with transparent wings and an upcurved spine at the end of the abdomen; BL 5–13 mm. Females are slightly smaller, antlike, somewhat flattened, and modified

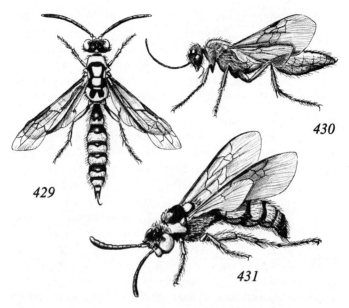

Figs. 429–431 Wasps: 429, Tiger Tiphiid, *Myzinum maculatum;* 430, Nocturnal Tiphiid, *Brachycistis* species; 431, Kingfisher Wasp, *Trielis alcione.*

with enlarged legs and spines for digging. RANGE: Sandy places in the deserts, Central Valley, and coastal areas.

VELVET ANTS
Family Mutillidae

Although not closely related to ants, mutillid females are wingless and densely covered with hair, somewhat resembling bulky ants as they erratically scurry about. Nocturnal species are pale or reddish brown with appressed hair forming velvety black or white bands; those active in the daytime are black wasps with brightly colored coats of long, white, red, or orange hair. Males resemble females but are winged. All species are parasitic on other insects, and females are most often seen in search of suitable hosts, mostly ground-nesting bees and wasps. They make a peculiar squeaking noise if held by the body, which is best done with forceps, for despite their pretty appearance, they can render a painful sting and are sometimes called "cow-killers or mule-killers." Probably the sting functions as defense against adult bees in their nests, because oviposition through the cell wall is not accompanied by a sting to paralyze the larval host. About 100 species of Mutillidae occur in California, representing more than a dozen genera, most of them nocturnal and restricted to the deserts.

Plate 16d. Velvet ants. *Dasymutilla.* Members of this genus are brightly colored and active by day. ADULTS: Medium-sized (BL 6–20 mm); rather robust, with a hard, dark integument. (**Plate 16d**) The **Red-haired Velvet Ant** *(D. coccineohirta)* is covered with yellow or orange to red hair. It is one of our commonest species, ranging widely in the Coast Ranges, Sierra Nevada, and Central Valley. **Sacken's Velvet Ant** *(D. sackeni)* is a large (BL 10–18 mm) species, covered with long, white or yellowish hair. It occurs widely in the Coast Ranges and Central Valley. A similar species, the **Glorious Velvet Ant** *(D. gloriosa),* has paler integument and is a desert species covered with long, white hair. Females resemble the bits of seed fluff from Creosote Bush *(Larrea)* that blows along the ground. The **Golden Velvet Ant** *(D. aureola)* has a squarish head, wider than the thorax, and is covered with golden yellow hair. It is common in the foothills of central California, ranging from northern to southern California.

Family Scoliidae

Scoliids resemble tiphiid wasps but are winged in both sexes and generally are much larger and relatively more robust. Like tiphiids, the scoliid wasps parasitize scarab beetle larvae in the soil. Four species occur in California, all of them most often seen at flowers in arid places.

431. Kingfisher Wasp. *Trielis alcione.* ADULT: Males are rather slender, with a black- and yellow-striped abdomen; females are large (BL 12–20 mm) with a broad, orange-banded abdomen and stout legs, adapted for digging. RANGE: Deserts, southern California, and the Central Valley, north to Antioch. The **Toltec Scoliid** *(Campsomeris tolteca)* is similar in color and range but is much larger (females BL 15–28 mm).

ANTS
Family Formicidae

There are many species of ants, with more than 400 recognized in the United States and about 200 in California, a few of them introduced from other parts of the world by man's activities. In all but a few parasitic forms, they are social and form colonies with a worker caste. Ants may be distinguished from all other insects, including their many mimics, by the elbowed antennae, an accentuated constriction (pedicel) between the thorax and abdomen, and by 1 or 2 abdominal segments greatly narrowed and bearing one or two bumps, or "nodes." There is much variation in body form, but usually the head is rectangular, wider than the narrow thorax, and the abdomen is large, globose. Males occur only during a brief mating season, when both sexes are winged. Otherwise, the colony consists of the queen and various types of wingless workers, modified females which do not lay eggs. Males have disproportionately large eyes, and worker ants in some groups have the eyes greatly reduced or absent.

Ants care for all stages of their progeny, feeding, cleaning, and transporting them from place to place as needed. Larvae are legless, sightless grubs and are fed by the workers, in many species only by regurgitation. There is a bewildering array of behavioral specialization and extensive literature on ant biology. The nest architecture and size, caste variation, foraging habits, etc. are all highly varied. Some species enslave other ants to carry out worker chores, while others are parasitic in nests of other ants. Still others are carnivorous and do not form permanent nests but bivouac in temporary quarters, wandering from place to place in long columns. Many ants are closely associated with certain plant-feeding insects that excrete a sweet liquid, "honeydew," especially certain lycaenid butterfly caterpillars, aphids, and scale insects. The honeydew is harvested by the ants, which in turn protect the donors from parasites and predators. In extreme examples, caterpillars are even carried back to the nest and raised there. Many ants are vegetarians, and some feed exclusively on fungi which they culture on small leaf fragments in special chambers in the nest.

In certain subfamilies a large and well-developed sting is present in females, including workers. Many ants can inflict a painful bite, whether or not they also sting in defense of the colony.

432. California Harvester Ant. *Pogonomyrmex californicus.* This is one of the common ants in California. Queens appear in spring and establish nests in sandy places. The colonies become quite large, with large or small craters at the entrance but without definite cleared areas around them. ADULT: Pale rust red in both queen (BL 8 mm) and workers (BL 5–6 mm). They forage for seeds during all but midday and frequently hold the abdomen erect. They bite and sting ferociously when disturbed. RANGE: Lower elevations, especially in the more arid areas. In southern California, workers are black and reddish in some colonies.

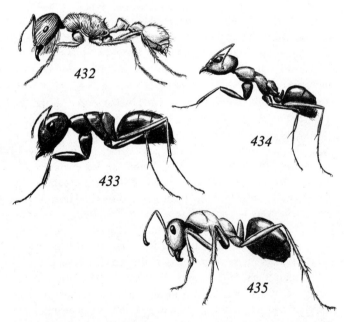

Figs. 432–435 Ants: 432, California Harvester Ant, *Pogonomyrmex californicus;* 433, Carpenter Ant, *Camponotus* species; 434, Argentine Ant, *Iridomyrmex humilis;* 435, Red Mound Ant, *Formica* species.

433. Carpenter ants. *Camponotus.* This genus includes the largest of our ants. They live in large colonies in soil or wood, often extending their galleries underground. Although formidable in appearance, they do not sting or bite readily. Foods include honeydew, sap, fruits, and flesh of other insects. The **Giant Carpenter Ant** *(C. laevigatus)* nests in fallen logs and stumps in forested areas. ADULTS: Shining black, sometimes with reddish brown legs; BL, workers 6–10 mm, queen 13–15 mm. RANGE: Middle to high elevations of the Coast Range and Sierra Nevada.

434. Argentine Ant. *Iridomyrmex humilis.* This native of South America was introduced into Louisiana about 1890 and spread westward, appearing in California in 1908. It forms large colonies of numerous nests in soil, debris, or rotting wood and destroys native ants in its territory. It is one of our worst household pests, eating all kinds of sweets and dead animal matter. The workers protect honeydew-secreting aphids and other insects from natural enemies. ADULT: Workers are small (BL 1–2 mm); dark brown with slender bodies and unusually long antennae. Queens are brown with a darker abdomen and silky hairs; BL 6 mm. RANGE: Warmer parts of the state, especially urban areas.

435. Red mound ants. *Formica.* These ants make nests in rotting wood or in soil, which they cover with conspicuous mounds of sticks and other vegetable detritus. ADULTS: Workers are rust red with the abdomen, and often parts of the legs, black; BL 3–8 mm. They actively guard and tend aphids for honeydew; foods are similar to those of carpenter ants. When the colony is disturbed, the workers swarm out in huge numbers, creating a rustling noise as they scramble over the mound, and biting any intruder foolish enough to loiter there. RANGE: Chaparral and sparsely forested areas of the Coast Ranges and Sierra Nevada.

Plate 16e. Honey ants. *Myrmecocystus.* Members of this genus inhabit arid and semiarid places. Colonies develop specialized workers, called ''repletes,'' which serve solely to store liquid nutrients during superabundance of foods in certain seasons. These individuals (**Plate 16e**) hang from gallery ceilings like bloated bags, filled to capacity with nectar from flowers, fruit juices, honeydew from aphids, etc. All species are general predator-scavengers. ADULTS: Females are large (BL 6–13 mm); workers and males smaller (BL 2–6 mm); pale yellowish to dark brownish, sometimes with paler appendages; densely covered with fine hair; petiole with one node. RANGE: About 18 species occur in the state, the most widely distributed being *M. testaceus*, which occurs in southern California, the Central Valley, and throughout the Great Basin. *M. mexicanus* (**Plate 16e**) is common in desert areas.

SPIDER WASPS
Family Pompilidae

Pompilids are slender wasps with long legs and antennae, recognizable in the field by their nervous-appearing behavior. They run erratically, darting under low clumps of vegetation or rocks, constantly twitching their wings. All species prey on spiders, which they quickly subdue by a powerful, paralyzing sting applied to the victim's underside. After capture the prey is usually dragged to a suitable site, where a simple burrow is dug with a terminal cell just large enough for the spider. The busy pompilid then drags the immobile arachnid into the chamber, deposits an egg on its abdomen, and closes the nest, obliterating all traces of it at the surface. A characteristic habitat is hunted, and spiders of similar size are taken by each pompilid; a single spider provides sufficient food for each offspring. The larvae are white, legless grubs which partially break down the spider's outside skeleton by using external enzymes, and feed by ingesting its fluids. There are about 130 species of spider wasps in California, most of them small and black or steel blue. They visit flowers of many kinds for nectar.

Plate 16f. Tarantula hawks. *Pepsis.* These are our largest and best known spider wasps. They attack and subdue the giant tarantulas and are often seen flying low over the landscape or visiting milkweed flowers in arid sites. ADULTS: Large (BL 14–48 mm); with relatively bulky, black or steel blue bodies

and broad, orange or black wings. RANGE: Deserts and foothills north in the Central Valley and east of the Sierra Nevada
to Lassen County. Among our most widespread species, *P.
thisbe* is the largest (female BL 32–44 mm); body deep steel
blue, wings bright orange. *P. mildei* is nearly as large and is
distinguished by its darker wings and bright orange antennae in
the male; *P. pallidolimbata* is smaller (female BL 18–32
mm), with the body black and the wings pale orange.

436. Forest Spider Wasp. *Priocnemis oregona.* Unlike most
pompilids, this species lives in shaded forest areas. It is commonly seen running about in leaf litter and on root-entangled
banks in search of burrowing spiders, which are probably en-
ticed from their burrows late in the afternoon. ADULT: BL
8–16 mm; entirely metallic blue-black except the abdomen
which is bright red. RANGE: Coast Ranges and Sierra Nevada
up to middle elevations. Common in the redwood zone from
Santa Cruz northward.

PAPER NEST WASPS, YELLOWJACKETS
Family Vespidae

The family Vespidae contains a diversity of yellow and
black wasps, including the social wasps, "yellowjackets" and
"hornets." Vespids are distinguishable by their thorax which
ends abruptly, hoodlike at the back of the head without a raised
rim or "collar." Nearly all species can fold their wings longitudinally. Solitary species build mud nests or dig tunnels in
the ground or use hollow stems in which they construct mud-
partitioned cells. Most hunt caterpillars, which they sting into
paralysis, storing several in a cell for each larva. Social species
build nests of paperlike material which they fabricate from
wood particles and saliva. The cells are honeycomblike in form
and are suspended upside down from an overhanging object
such as eaves, or in natural cavities in the ground or in standing
dead trees. Or they may be hung in trees and covered with
layers of the wasp-paper. These nests are used for one season.
Construction is initiated in spring by an overwintering female,
or queen. She lays eggs, and produces nonreproductive workers which in some genera are smaller than the queen. They

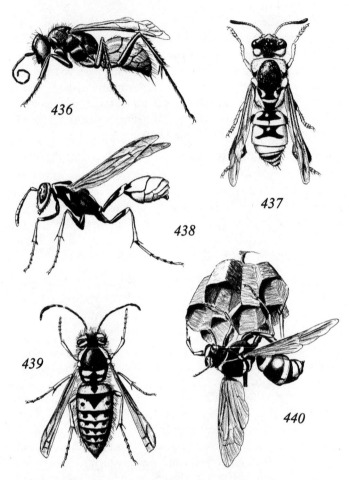

Figs. 436–440 Wasps: 436, Forest Spider Wasp, *Priocnemis oregona;* 437, Common Eumenid, *Euodynerus annulatum;* 438, Yellow-Legged Paper Wasp, *Mischocyttaris flavitarsis;* 439, Pennsylvania Yellowjacket, *Vespula pensylvanica;* 440, Golden Paper Wasp, *Polistes fuscatus.*

gather miscellaneous animal matter to feed the larvae. Males are usually produced toward the end of the summer.

More than 100 species occur in California, all but 15 of them solitary, although the social species are all larger and more conspicuous wasps.

437. Common Eumenid Wasp. *Euodynerus annulatum*. This is one of the commonest of the solitary vespids in California. Females make vertical burrows in the ground, topped by a mud chimney, and provision them with caterpillars. ADULT: BL 7–14 mm; stout, with a globular abdomen nearly as broad as the thorax; yellow with variable amounts of black markings. RANGE: Widespread in arid areas, including the deserts and foothills throughout the state.

438. Yellow-legged Paper Wasp. *Mischocyttarus flavitarsis*. This species builds nests similar to those of *Polistes,* and the larvae are fed caterpillars and other soft insects. The adults feed at flowers or fruits and often enter buildings to overwinter, appearing at the windows in spring. ADULT: BL 16–20 mm; similar to *Polistes* but with the abdomen smaller, globose, attached to the thorax by a long, slender, stalklike segment; thorax mostly black, abdomen and legs yellow with black markings; queens and workers alike. RANGE: Widely distributed at lower to middle elevations but absent in the deserts.

439. Yellowjackets. *Vespula*. Members of this genus are the most abundant and troublesome wasps in California. They build nests of open-faced paper combs which are tended by increasing numbers of workers as the season progresses. The voracious workers attack everything in the vicinity, from resting insects to pieces of hamburger on the picnic table. They are vicious defenders of the nest and deliver extremely painful stings when disturbed, attracting their sisters with an aggregation chemical. ADULTS: Stout, with the head, thorax, and abdomen of about equal width, the abdomen appearing to be broadly attached to the thorax; yellow with black markings; BL, queens 16–19 mm, workers 11–15 mm. RANGE: Throughout coastal areas, foothills, and mountains. There are 2 subgenera: *Vespula* in the strict sense, with about 6 species in California, build uncovered, flat-surfaced combs in existing cavities, such as in rodent burrows, underground in jumbled debris, and in hollow logs. The colonies may become very large by late summer or fall. The **(439) Pennsylvania Yellowjacket** *(V. pensylvanica)* is one of the most abundant and widespread species. Members of the other subgenus, *Dolichovespula,* are usually less numerous because colonies tend not to last beyond mid-summer. They build aerial nests, suspended from tree branches or in bushes. The combs are turned up at the margins and are surrounded by protective layers of paper sheeting. Each successive, larger comb is suspended below the preceding, and the nests become cone- or football-shaped, with the entrance near the bottom. Among the 3 species in California, the most conspicuous is the **White-faced Hornet** *(V. maculata)* which is much larger than the common yellowjackets (BL 17–22 mm), has whitish rather than bright yellow markings. It is restricted to the northern half of the state.

440. Golden Paper Wasp. *Polistes fuscatus.* This species builds small, open-faced combs of 6-sided cells, suspended by a stalk under bridges, eaves of buildings, etc. Although capable of stinging, the workers are not as easily disturbed near the nests as are yellowjackets. ADULT: BL 11–19 mm; queens and workers of equal size; slender; with a narrowly constricted waist; thorax black with yellow markings, abdomen mostly yellow. RANGE: Widespread at lower elevations; the markings are brick red instead of black in the deserts and other arid parts of southern California. Nests are provisioned with various food of animal origin, such as caterpillars, which are skinned by the adult wasps. About 6 similar species of *Polistes* occur in the state.

DIGGER WASPS, THREAD-WAISTED WASPS
Family Sphecidae

Sphecid wasps make up one of the most interesting insect families because of their amazing diversity in nesting behavior. Each genus hunts specific kinds of insects or spiders, and there are remarkable specializations in prey transportation, burrow architecture, and the manner in which nests are maintained. Many books have been written on the subject, yet the field of study remains a fertile one, with habits of some genera and species unknown or not fully investigated.

These wasps vary greatly in form and size, from slender and thread-waisted to rather stout. They are distinguishable from vespid and pompilid wasps by having a visibly constricted neck and a raised "collar" around the rim of the thorax. Most are equipped with digging combs on the forelegs, and they nest in the ground; but some use hollow stems or abandoned burrows of other insects, while a few construct free mud cells. Specialized hunters prey on orb-weaving or ground-inhabiting spiders, crickets of particular kinds, lepidopterous caterpillars, adult insects of nearly every order, even aphids or thrips, and some sphecids use the nests and provisions of other sphecids. In most genera the cells are provisioned, an egg laid in each, and they are closed off prior to larval feeding; but in a few the cells are provisioned progressively as the larva grows. In the extreme case, more than one nest is maintained simultaneously by the parent.

More than 350 described species occur in California, occupying every conceivable habitat, from sea coast dunes to timberline. A great many are small wasps, members of genera not included in the following examples.

441. Yellow and Black Mud Dauber. *Sceliphron caementarium.* This is one of our commonest wasps. Each female builds mud nests of cells, laid side by side in series usually of 2–6, the whole covered by a continuous layer of mud. Nests are placed on undersides of rocks, under eaves, etc., then provisioned with spiders, primarily crab-spiders and others that live on plants. ADULT: BL 14–28 mm; black with collar, portions of the thorax, legs, and the conspicuously thread-waisted abdomen yellow. RANGE: Widely distributed at low to middle elevations.

442. Golden Digger Wasp. *Sphex ichneumonia.* This species usually constructs a simple burrow in heavy soil and provisions it with a single, large katydid nymph which has been dragged along the ground by the backward-walking wasp. ADULT: Large (BL 15–27 mm); thorax gray with dense, crinkly hair; abdomen bright reddish orange, covered with golden pile. RANGE: Widespread in arid parts of the state, deserts, southern California, and Central Valley.

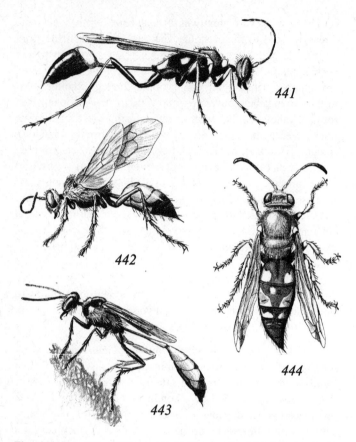

Figs. 441–444 Sphecid Wasps: 441, Yellow And Black Mud Dauber, *Sceliphron caementarium;* 442, Golden Digger Wasp, *Sphex ichneumonia;* 443, Thread-Waisted Digger Wasp, *Ammophila* species; 444, Western Cicada Killer, *Sphecius convallis.*

443. Thread-waisted digger wasps. *Ammophila.* This genus is of great interest because the whole range of simple to complex prey provisioning is shown. The nest always is a simple, vertical burrow with one cell, and during its provisioning the female uses a rock which exactly fits the entrance, as a temporary door. At final closing she often holds the rock in her mandibles and tamps the fill with it. All species prey on caterpillars, some using large ones like sphinx moth larvae, providing one to a cell after laboriously dragging it from the point of capture to the nest. Other *Ammophila* provide several smaller caterpillars, such as inchworms, which they can carry in flight one at a time. Still others provision progressively; the mother decides each morning, based on the size of her offspring, how many prey to bring in. Individual females of some *Ammophila* even maintain 2 or 3 nests at once, each larva in a different stage of development. ADULTS: BL 8–38 mm; very slender, abdomen formed into a long thread-waist with a small, oval enlargement at its tip; thorax usually black, marked with silver, the abdomen gray or silvery with the enlarged tip orange or reddish. RANGE: Widely distributed, especially in sandy areas. More than 20 species occur in the state.

444. Western Cicada Killer. *Sphecius convallis.* The largest wasp in California, this species hunts large game, adult cicadas. The paralyzed victim is carried in flight, held beneath the wasp's body by the middle- and hindlegs, the latter specially modified to hold the cicada. Nests are dug in soft ground, are simple with one cell, and each provisioned with one cicada. ADULT: BL 15–35 mm; robust, with a broad thorax and abdomen nearly as broad, without a thread-waist; thorax black, abdomen rust brown with yellow bands. RANGE: Central Valley and deserts.

Plate 16g. Sand wasps. *Bembix.* Familiar to any naturalist interested in sand dunes, the industrious females of this genus are commonly seen rapidly kicking sand from their large tunnels, which are often aggregated in small colonies. Nests of some species are complex feats of architecture, with a great deal of variation from one species to another, often involving several cells. Females of some species progressively provision as the offspring grows. All *Bembix* hunt adult Diptera, which they evidently catch on the wing, and a variety of flies may be taken by a given female. ADULTS: BL 12–22 mm; stout-bodied, rather robust wasps with the abdomen appearing to be broadly attached to the thorax; gray or black with pale to bright yellow markings. RANGE: Widespread, mostly at lower elevations in sandy areas, such as beach and desert dunes and riverbanks. More than a dozen species occur in California.

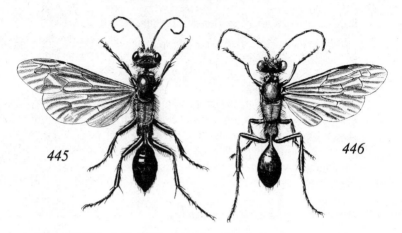

Figs. 445–446 Sphecid Wasps: 445, Green Cricket Hunter, *Chlorion aerarium;* 446, Blue Mud Wasp, *Chalybion californicum.*

445. Green Cricket Hunter. *Chlorion aerarium.* One of our most beautiful wasps, this species flushes crickets from their daytime hiding places and buries them in a simple nest in the ground. ADULT: Large (BL 16–30 mm); slender with a thread-waist; metallic bright blue-green or blue with dark, violet-tinged wings. RANGE: Deserts and Central Valley.

446. Blue Mud Wasp. *Chalybion californicum.* This wasp is in a sense a "poor relative" of the (**441**) Yellow and Black Mud Dauber, for the female *Chalybion* uses abandoned cells of *Sceliphron* or other similar niches, carrying water to soften the old adobe for patching and sealing. She does not carry mud and seems unable to do her own architecture. The Blue Mud Wasp hunts primarily on the ground, preying mainly on Black Widow spiders and other members of its family. ADULT: BL 10–23 mm; metallic blue, blue green, or bluish black; slightly built but with a shorter thread-waist than in *Sceliphron*. RANGE: Widespread: Sierra Nevada, Coast Ranges, and Central Valley to the desert margins.

BEES
Superfamily Apoidea

When most people mention "bees" they are referring to domesticated Honey Bees. There are, however, over 1,000 species of native bees in California. Only a few are social, including the Honey Bee (an introduced species from the Old World) and the bumble bees. Both of these are members of the family Apidae, while the vast majority of bees are solitary in habits and are members of several other families. Like wasps, most bees dig nests in the ground, some burrow into wood, soft stems, etc., and a few construct nests of resin or mud. In contrast to wasps, bees collect pollen as provision for their larvae, and various specializations have been developed for pollen-carrying. Many bee species collect pollen from only one or a few plant species.

Bee larvae are legless, white grubs, living entirely within a cell, fed pollen provided by the solitary mother or in the case of social species by the workers. A few genera have evolved a parasitic mode of life, laying their eggs in nests of other bees, in which the occupant is killed at an early stage and the food supply eaten.

Compared to wasps, bees in general are more robust, hairy insects, having branched hairs, especially on thorax. The abdomen is broadly joined to the thorax, and the hindlegs are broadened in the fifth-from-last segment, rather than narrowed as in wasps.

Family Andrenidae

447. Common burrowing bees. *Andrena*. One of the largest genera of animals, *Andrena* is represented by more than 150 species in California. Burrows are made in the ground, usually consisting of a vertical tunnel with lateral branches to each cell. ADULTS: Small to medium-sized (BL 6–15 mm); thorax densely hairy; abdomen usually black or metallic blue, often with narrow, discrete, pale hair bands; females with long hair brushes on the hindlegs; males lack the brushes and have narrower, pointed abdomens. RANGE: Virtually throughout the state. Most species fly in spring, visiting willow, and other early-flowering plants. A few of our larger species are the following: The (**447**) **Orange-banded Andrena** *(A. prunorum)* is one of the most abundant and widespread species in California, especially in the deserts and southern mountains. ADULT: BL 8–15 mm; males slender with dense, whitish hair and a yellow face; females more robust with golden ochreous

Figs. 447–450 Bees: 447, Orange-Banded Andrena, *Andrena prunorum;* 448, Resin Bee, *Anthidium* species; 449, Short Leafcutter Bee, *Megachile brevis;* 450, Metallic Leafcutter Bee, *Osmia* species.

hair and the facial yellow variable, restricted; wings in both sexes brownish, shaded darker towards the tips; abdomen black with whitish hair bands, often the basal segments orange, superficially resembling the Honey Bee. Visits a wide variety of flowers. The **Greenish Burrowing Bee** *(A. cuneilabris)* is one of the commonest andrenids in central California. It is often seen in early spring visiting buttercups *(Ranunculus).* ADULT: BL 9–12 mm; metallic bluish green with densely hairy head and thorax. RANGE: Foothills of the Coast Ranges and Sierra Nevada. **Black bees** *(Andrena [Onagrandrena]).* Several species of this subgenus fly at dusk or before sunrise in order to gather pollen from evening primroses *(Oenothera);* the females have specially modified hair brushes to accommodate the cobwebby pollen, and they are the only pollinators of some evening primroses. ADULTS: Moderately large (BL 9–15 mm); entirely black, densely covered with black hair in the female; males are slender, black with whitish hair on the thorax in some species. RANGE: Sandy areas of the deserts, Central Valley, and seacoast dunes.

Family Megachilidae

Members of the family Megachilidae differ from all other bees in having the pollen-collecting brush on the underside of the abdomen, rather than on the legs. Most species cut out circular or oval discs from leaves, or use other plant materials to build the walls of the nest cells. There are more than 250 species described in California.

448. Mason or **resin bees.** *Anthidium, Dianthidium.* The rapid flight accompanied by a high-pitched whine makes bees of these and related genera distinctive. They are solitary and build cells of plant resins, plant fibers, mud, or pebbles cemented by resins. Many plants are used as pollen sources, but *Phacelia* and deerweed *(Lotus)* are most commonly visited. Nests are constructed in abandoned insect or spider burrows in the ground or in wood, and a few species dig burrows in sand. *Anthidium* females gather plant fibers with which they line the cells, while *Dianthidium* use fluffy down scraped from plants. ADULTS: BL *Anthidium* 8–12 mm, *Dianthidium* 6–12 mm; stout, cylindrical, with the abdomen curled downward towards the tip; gray or black, marked with yellow spots, distinctively arranged in rows along the abdominal segments. RANGE: The usually larger *Anthidium* are primarily montane, ranging along the Sierra Nevada (19 species including some restricted to high elevations) and Coast Ranges, although a few are desert species; *Dianthidium* (13 species) tend to occur in more arid places, in the deserts and Central Valley or foothills.

449. Common leafcutter bees. *Megachile.* ADULTS: BL 7–17 mm; generally cylindrical, moderately robust, abdomen with a squared-off appearance; gray or black, usually with distinct, whitish hair bands around the abdomen. More than 60 species occur in California, many of them preferring flowers of the sunflower family for pollen. Nesting occurs in pre-existing holes in the ground or in wood. The **Angelic Leafcutter Bee** *(M. angelica)* is one of our smaller but most common species. Adults are blackish with moderately distinct, thin abdominal hair bands; BL 7–11 mm. It is common during the summer throughout mountains and foothills from the Transverse Range northward, visiting a wide variety of flowers. The **Alfalfa Leafcutter Bee** *(M. rotundata)* is introduced from Europe. It has been cultivated as a pollinator of alfalfa, which is not efficiently pollinated by Honey Bees, by providing wooden racks with holes of an appropriate diameter as nest sites. Producers of alfalfa seed are said to have increased yields approximately tenfold through use of this bee. (**449**) The **Short Leafcutter Bee** *(M. brevis)* is a common native species. ADULT: Small (BL 7–11 mm); blackish with thin, white abdominal hair bands. Visits a diversity of flowers, especially sunflowers. RANGE: Low to moderate elevations over most of the state except the deserts.

450. Metallic leafcutter bees. *Osmia.* The brightly colored members of this genus are among our most common mountain bees. They visit a wide range of flowers and nest in abandoned beetle burrows in logs, or in other holes. ADULTS: BL 5–15 mm; stout, cylindrical with round abdomens; bright metallic green, blue, or golden. RANGE: Throughout foothill and mountainous parts of the state. More than 75 species have been described from California, but their taxonomic relationships and biologies are poorly known.

Family Halictidae

451. Alkali Bee. *Nomia melanderi.* This species is solitary but nests in large aggregations and is encouraged as a pollinator. The females prefer lowland, alkaline soils, such as waste areas around alfalfa fields, and are more efficient pollinators of alfalfa than Honey Bees are. ADULT: Moderately large (BL 9–14 mm); robust with smooth abdomen, having distinctive whitish bands tinged with abalone shell colors. Females collect pollen from a variety of plants. RANGE: Lower elevation sites in the Central Valley south through the deserts.

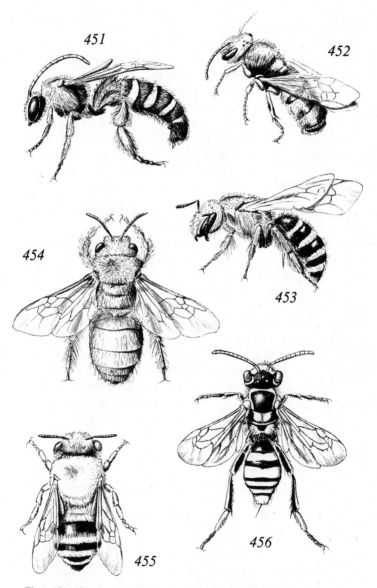

Figs. 451–456 Bees: 451, Alkali Bee, *Nomia melanderi;* 452, Metallic Sweat Bee, *Agapostemon* species; 453, Semisocial Sweat Bee, *Halictus* species; 454, Squash Bee, *Peponapis pruinosa;* 455, Urban Anthophora, *Anthophora urbana;* 456, Edwards' Cuckoo Bee, *Nomada edwardsii.*

452. Metallic sweat bees. *Agapostemon.* Bees of this genus resemble cuckoo wasps in the metallic green color, but lack the shiny, concave underside of the abdomen that characterizes chrysidids. ADULTS: Females have broad, pollen-collecting brushes (scopae) on the hind legs; entirely metallic green or with dark-banded abdomen. Males (**452**) are slender; metallic green, abdomen yellow with black bands; hind legs lack the brushes; BL 7–14 mm. A variety of flowers is visited by any given population. RANGE: Throughout low to middle elevations of the state including the deserts. About 6 species occur in California.

453. Semisocial sweat bees. *Halictus, Lasioglossum.* Sweat bees received their common name because they often are attracted to perspiring skin, for example at swimming pools, and sometimes give a sharp but harmless sting. These are among California's most common bees because there are more than 70 species, and many are somewhat social in habits, producing a large number of worker females which are similar to the queen. There is wide variation, from solitary species to organized social colonies in which the "queen" ceases provisioning after the first brood. In some species workers may produce unfertilized eggs that result in males. The nests are complex burrows in the ground and, even in species that provision in a solitary fashion, many females may use the same entrance. ADULTS: Small to medium-sized (BL 3–13 mm); dull metallic gray, brown, or blackish, often with distinct, pale hair bands on the abdomen. In females the last visible abdominal segment is split longitudinally on the upper side. Many kinds of flowers are used by each colony. RANGE: Virtually throughout the state except at the highest elevations.

Family Anthophoridae

The Anthophoridae includes more than 300 species in California. Most are robust, hairy bees which dig solitary, often very deep, burrows in the ground. Members of a number of genera, however, are slender, hairless species which parasitize the nests of other bees, either those of other families or of particular pollen-collecting genera of Anthophoridae.

454. Squash Bee. *Peponapis pruinosa.* This species collects pollen only from plants of the squash family, and throughout most of California pollination of gourds, squashes, and pumpkins is largely dependent upon this bee. ADULT: Males are medium-sized (BL 10–13 mm); gray with tan hair on thorax and abdominal bands; females are more robust (BL 11–14 mm) with more extensive, ochreous hairiness. RANGE: Throughout squash-growing areas of the state, having moved northward with cultivation following domestication of squashes in Mexico. Pollen-collecting takes place primarily before sunrise.

455. Urban Anthophora. *Anthophora urbana.* This is one of the most common native solitary bees in California. ADULT: Medium-sized (BL 10–13 mm); with distinct, pale gray hair bands on the abdomen. RANGE: Widespread in foothills and middle elevations of the state; seen most often in late spring. Another species, **Edwards' Anthophora** *(A. edwardsii),* is more prevalent in early spring when few other bees are flying. Many plants are visited, but early-flowering kinds such as manzanita *(Arctostaphylos)* are most often used. ADULT: Large, robust (BL 12–18 mm); thorax covered with pale gray pile; abdomen black without pale hair bands; males with a white or yellow facial patch. RANGE: Throughout lower elevations and foothills. More than 30 other species of *Anthophora* occur in California. Members of the related genus *Synalonia* are also among our most common bees visiting spring wildflowers, especially fiddleneck *(Amsinckia).* ADULTS: Moderately large (BL 8–15 mm); stout with densely hairy thorax and shiny, round abdomen, sometimes with pale hair bands; males have long antennae, nearly as long as the body, and a yellow patch on the face. RANGE: Throughout coastal, foothill, and low-elevation areas of California.

456. Cuckoo bees. *Nomada.* The slender members of this genus resemble wasps. They are commonly seen hovering low over the ground in search of other bees' burrows (especially Andrenidae) in which *Nomada* deposits her eggs. ADULTS: Slender-waisted with a globular thorax and pointed abdomen, which often is held upright; wings characteristically margined with brown. **(456) Edwards' Cuckoo Bee** *(N. edwardsii)* is a large (BL 9–12 mm), yellow and black species resembling a vespid wasp. It occurs in foothill and middle-elevation areas during late spring. The common **Red Cuckoo Bee** *(N. angelarum)* is most often seen early in the season, hovering around nest sites or visiting manzanita and other early spring flowers. It is a brick reddish species with darker thorax and brownish wings. It is common and widespread, but its biology and taxonomy are not well known, as is true of all the 60+ species of *Nomada* described from California.

457. Carpenter bees. *Xylocopa.* The largest bees in California, members of this genus construct long burrows in standing dead trees, structural timbers, fenceposts, etc. Each burrow consists of a linear series of cells partitioned with discs made of wood chips. Although the bees congregate during nesting and a single entrance may lead to several burrows, there is no true social behavior. ADULTS: Large and robust (BL 12–26 mm); deep metallic blue, black, or blond in the males of some species. Females have a dense brush of hairs on the hindlegs for gathering pollen. RANGE: Throughout the state except at high elevations. The **Mountain Carpenter Bee** *(X. tabaniformis)* is a small species (BL 12–18 mm); black in both sexes with variable amounts of yellowish hair on the thorax of the male. It occurs in the Coast Ranges and Sierra Nevada using dead Incense Cedar and redwood structural timbers for nesting.

Figs. 457–458 Bees: 457, Valley Carpenter Bee, *Xylocopa varipuncta;* 458, Honey Bee, *Apis mellifera.*

(457) **Valley Carpenter Bee** *(X. varipuncta)* has metallic black females and pale buff males; BL 18–26 mm. It is the common species in southern California and occurs northward through the Central Valley, using old hardwoods and telephone poles as burrow sites. The commonest species, the **California Carpenter Bee** *(X. californica)*, is shining blue or greenish blue with dark wings; BL 17–26 mm; it occurs in the deserts and southern mountains where it uses the dry stalks of Yucca *(Yucca whipplei)* and agave for nesting. Northward, this bee occurs in the foothills and mountains and nests in Incense Cedar and redwood timbers.

Family Apidae

Plate 16h. Bumble bees. *Bombus*. This genus contains the familiar medium-sized to large, densely hairy, yellow and black bees that are commonly seen in gardens, weedy areas, and natural habitats. They live in social communities consisting of sexual females or queens, males or drones, and asexual females or workers. The large queen overwinters and in spring seeks a secluded hollow such as an abandoned rodent's nest to begin the colony. She then gathers pollen and nectar and lays eggs. The resulting offspring are the much smaller workers which take over gathering of pollen and nectar and the construction of brood cells. In the fall, males as well as new queens are produced from unfertilized eggs, and mating takes place prior to hibernation. Many kinds of flowers are visited, but because of their size, bumble bees are able to use and pollinate clover, alfalfa, and other flowers that Honey Bees and other short-tongued bees cannot. Bumble bees, because they self-regulate their body temperature and are active at colder air temperatures, more efficiently use early-spring and high-elevation plants, and are active early and late in the day and in cloudy weather. ADULTS: Queens are large, workers resemble the queens but are smaller, and both have pollen baskets formed of broadened, concave, hindleg segments marginally fringed with long hairs. Most species are black with yellow hair bands, while some are largely yellow and some high-elevation species have a reddish abdominal band as well. RANGE: Throughout the state except in the deserts. The

commonest species at lower elevations and around cities is the
(**Plate 16h**) **Yellow-faced Bumble Bee** *(B. vosnesenskii)*,
which is black, with bright yellow face, thorax in front of the
wings, and a narrow band on the rear of the abdomen; BL,
queens 15–22 mm, workers 10–17 mm. Queens of the **Sonoran Bumble Bee** *(B. sonorus)* are among the largest bumble
bees in California (BL to 24 mm). They appear in early spring
in southern California and the Central Valley and are densely
covered with deep yellow pile, with a black band across the
thorax, and the end of the abdomen black. Workers are similar
but smaller, varying greatly in size (BL 10–20 mm).

458. Honey Bee. *Apis mellifera.* Probably the world's best
known insect, the Honey Bee is not native to North America. It
was introduced into California about 1850 when hives were
shipped around the Horn, and now it is naturalized virtually
everywhere. The morphology and habits of this insect have
been studied from Renaissance times onward. The social or-
ganization is highly developed with 3 forms: the queen, workers, and males or drones. The queen has a long abdomen,
performs no nest functions other than egg-laying, and may live
several seasons. There may be 50,000 or more workers in a
colony, and each lives only a few weeks, or a few months in
the case of winter workers. Drones are larger and stouter than
workers, with large eyes that meet at the top of the head. In all
castes the thorax is hairy, brownish, the abdomen orange with
variable extent of black banding; BL workers 10–15 mm.

When the population increases beyond the capacity of the
hive, a new queen is produced by special feeding, and a
swarm, consisting of the old queen and a large number of
workers, is emitted to colonize elsewhere. The virgin queen
takes a nuptial flight, followed by a number of males. Copula-
tion occurs in mid-air, and the fertilized queen returns to the
nest. At the end of summer, the workers eject the males from
the nest since the colony will not develop swarms again until
the following year; the males, with no further function in the
life of the community, die.

The honeycomb is the structural base of the colony. It is
composed largely of hexagonal cells, placed back to back, and

is constructed of wax by younger workers. The cells' sizes vary according to the caste destined to occupy them. Development requires about 16 days for workers and 24 days for drones; the larvae feed on a mixture of honey and digested pollen provided by the workers.

Honey Bees visit a variety of flowers. Communication among them has been extensively studied; workers carry out a complex dance on the comb which provides their sisters with cues as to direction and distance to nectar and pollen sources.

LEARNING MORE ABOUT INSECTS

This book will serve as a guide to information about only a small proportion of California's insect species. Undoubtedly readers will have many questions concerning the biologies of these and other species, or other aspects of entomology. Many specialized sources of information are cited for the individual orders, and there are general references available. Some of the major areas are listed below.

LITERATURE SOURCES
General and Identification

Borror, D., D. DeLong, and C.A. Triplehorn. 1976. *An Introduction to the Study of Insects,* 4th ed. New York: Holt, Rinehart and Winston.

Borror, D., and R. White. 1970. *A Field Guide to the Insects of America North of Mexico.* Boston: Houghton Mifflin Co.

Brues, C., A. Melander, and F. Carpenter. 1954. *Classification of Insects,* 2nd ed. Harvard, Bulletin Museum Comparative Zoology, Vol. 108.

Bulletin of the California Insect Survey, Vols. 1–22. Berkeley and Los Angeles: University of California Press (ongoing synopses of identification, distribution, and biology of selected taxa).

Daly, H.V., J.T. Doyen, and P.R. Ehrlich. 1978. *An Introduction to Insect Biology and Diversity.* New York: McGraw-Hill Book Co.

Essig, E.O. 1958. *Insects and Mites of Western North America,* 2nd ed. New York: Macmillan.

Farb, P., and editors of Life. 1962. *The Insects, Life Nature Library.* New York: Time Inc.

Evans, H.E. 1966. *Life on a Little Known Planet.* New York: E.P. Dutton Co.

Hogue, C.L. 1974. *The Insects of the Los Angeles Basin.* Nat. Hist. Mus. L.A. Co., Sci. Series.

Imms, A. 1957. *A General Textbook of Entomology,* 9th ed. London: Methuen and Co.

Jaques, H. 1947. *How to Know the Insects,* 2nd ed. Dubuque: W.C. Brown Co.

Klots, A., and E. Klots. 1959. *Living Insects of the World.* Garden City: Doubleday.

Ross, H. 1965. *A Textbook of Entomology,* 3rd ed. New York: Wiley and Sons, Inc.

Swan, L.A., and C.S. Papp. 1972. *The Common Insects of North America*. New York: Harper and Row.

Wigglesworth, V. 1964. *The Life of Insects*. Cleveland: World Publishing Co.

Zim, H., and C. Cottam. 1951. *Insects. Golden Nature Guide*. New York: Simon and Schuster.

Special Topics

Chapman, R. 1971. *The Insects, Structure and Function*, 2nd ed. New York: American Elsevier Publishing Co., Inc.

Chu, H. 1949. *How to Know the Immature Insects*. Dubuque: W.C. Brown Co.

Clausen, C.P. 1940. *Entomophagous Insects*. New York: McGraw-Hill Book Co.

Cooper, E. 1963. *Insects and Plants, the Amazing Partnership*. New York: Harcourt, Brace, and World.

Felt, E.P. 1940. *Plant Galls and Gall Makers*. Ithaca, New York: Comstock Publishing Co.

Furniss, R. L. and V. M. Carolin. 1977. *Western Forest Insects*. Washington, D.C.: Govt. Printing Off.; 654 pp.

Merritt, R.W., and K.W. Cummins (eds.) 1978. *An Introduction to the Aquatic Insects of North America*. Dubuque: Kendall-Hunt Co.

Michener, C., and M. Michener. 1951. *American Social Insects*. Toronto: Van Nostrand, Inc.

Peterson, A. 1967. *Larvae of Insects*, 6th ed. Ann Arbor: Edwards Bros. 2 vols.

Price, P.W. 1975. *Insect Ecology*. New York: John Wiley and Sons.

Snodgrass, R. 1935. *Principles of Insect Morphology*. New York: McGraw-Hill Book Co.

Sterling, D. 1954. *Insects and the Homes They Build*. Garden City: Doubleday.

Usinger, R. (ed.), 1956. *Aquatic Insects of California*. Berkeley: University of California Press.

Wigglesworth, V. 1973. *The Principles of Insect Physiology*, 7th ed. London: Methuen.

Wilson, E.O. 1971. *Insect Societies*. Cambridge: Harvard University Press.

Insects of Economic Importance

Davidson, R.H., and L.M. Peairs. 1966. *Insect Pests of Farm, Garden and Orchard*, 6th ed. New York: J. Wiley and Sons.

DeBach, P. 1974. *Biological Control by Natural Enemies*. New York: Cambridge University Press.

Ebeling, W. 1959. *Subtropical Fruit Pests*. Berkeley: University of California Division Agricultural Sciences.

Ebeling, W. 1975. *Urban Entomology*. Berkeley: University of California Division Agricultural Sciences.

Huffaker, C.B. (ed). 1971. *Biological Control.* New York: Plenum Press.

James, M.T. 1969. *Herms's Medical Entomology,* 6th ed. New York: Macmillan.

Mallis, A. 1967. *Handbook of Pest Control,* 5th ed. New York: MacNair-Dorland Co.

Metcalf, C., W. Flint, and R. L. Metcalf. 1962. *Destructive and Useful Insects: Their Habits and Control,* 4th ed. New York: McGraw-Hill Book Co.

Pfadt, R.E. (ed). 1971. *Fundamentals of Applied Entomology,* 2nd ed. New York: Macmillan.

Further sources include the innumerable circulars, pamphlets, bulletins, etc. published by the U.S. Department of Agriculture (available from Government Printing Office, Washington, D.C.), California Agricultural Experiment Station, Extension Service (University of California, Berkeley), state, county, and city agricultural and public health offices.

Periodicals

The following journals and newsletters feature articles of interest to students of the California insect fauna.

California Vector Views. Published monthly by the California State Department of Health Services, Vector Biology, and Control Section, 2151 Berkeley Way, Berkeley, Calif. 94704.

Coleopterists' Bulletin. Published quarterly by the Coleopterists' Society. Editor, Dr. D. R. Whitehead, c/o Department of Entomology, Stop 168, Smithsonian Institution, Washington, D.C. 20560.

Entomological Society of America, Bulletin and 6 issues per year of Annals, Environmental Entomology, and Journal of Economic Entomology. Executive Director, 4603 Calvert Rd., College Park, Md. 20740.

Journal of Research on the Lepidoptera. Published quarterly by The Lepidoptera Research Foundation, Inc., 2559 Puesta Del Sol Rd., Santa Barbara, Calif. 93105.

Journal of The Lepidopterists' Society. Published quarterly by the Lepidopterists' Society. J. P. Donahue, Secretary, Dept. Entomology, Los Angeles County Museum of Natural History, Exposition Blvd., Los Angeles, Ca. 90007.

Bulletin of the Oregon Entomological Society and Proceedings of the Washington State Entomological Society. Published by the Ore. Ent. Soc., c/o Entomology Department, 2046 Cordley Hall, Oregon State University, Corvallis, Ore. 97331.

Odonatologica. Published quarterly by the Societas Internationalis Odonatologica, Dr. K. J. Tennessen, secretary, 1949 Hickory Ave, Florence, Alabama 35630, or Dr. W. J. Westfall, Jr., Department of Zoology, University of Florida, Gainesville, Fla. 32611.

Pan-Pacific Entomologist. Published quarterly by the Pacific Coast Entomological Society, c/o Department of Entomology, California Academy of Sciences, Golden Gate Park, San Francisco, Calif. 94118.

Insect Study

Barker, P.C., et al. 1977. *A Study of Insects*. 4-H Entomology project. Div. Agric. Sci., Univ. of Calif. Leaflet 2949; 41 pp. (available from Agriculture Extension Service, University Hall, Univ. of Calif., Berkeley 94720 and local farm advisor's office).

Brown, V. 1968. *How to Follow the Adventures of Insects*. Healdsburg, California: Naturegraph.

Cummins, K., L. Miller, N. Smith, and R. Fox, 1964. *Experimental Entomology*. New York: Reinhold.

Kalmus, H. 1960. *101 Simple Experiments with Insects*. Garden City: Doubleday.

Oldroyd, H. 1958. *Collecting, Preserving, and Studying Insects*. New York: Macmillan.

Siverly, R. 1962. *Rearing Insects in Schools*. Dubuque: W.C. Brown, Co.

Tweedie, M. 1968. *Pleasure from Insects*. New York: Taplinger.

Ward's Natural Science Establishment, Inc. 1958. *How to Make an Insect Collection,* revised. Rochester, N.Y.: Ward's N.S.E., Inc.

MAKING AN INSECT COLLECTION

The study of insects is called entomology, and insect collecting is not only an excellent learning experience for the serious entomology student, but may be an exciting and satisfying hobby for anyone. Owing to the enormous variety, new additions to the collection are possible almost endlessly.

A basic difference between populations of birds and mammals and those of invertebrates including insects is the number and care of offspring produced by each female. Generally mammals and birds produce only a few young in each litter or clutch, and considerable parental attention is necessary to raise them to maturity. By contrast, most insects average 200 or more eggs per female, and in nearly all kinds, the immatures are left to provide for themselves. This means that in order to maintain stable population numbers, 198 young from every complement of 200 eggs must die before reaching reproductive maturity. Also, the mother insect need not live longer than production of the eggs requires, usually only a few days. In a mammal or bird averaging 4 young, the female must survive the feeding period of the offspring, and she can afford to lose only 2 if a stable population is to be maintained. Moreover, when abnormally large numbers of an insect are killed through some catastrophe, the percent of predation on the survivors' offspring is lower because prey are more difficult to locate in less dense colonies, so that losses of adult insects are in part counterbalanced by greater survival rates of ensuing immatures.

As a result, in contrast to birds and mammals, there is no evidence that collecting adult insects, even in relatively small populations, affects the overall numbers in succeeding generations. It is obvious that man's effects, such as urban sprawl, agriculture, transport of weeds, and mining of sand dunes, are responsible for wholesale destruction of habitats and the insect populations inhabiting them, especially on a local scale. But insects are amazingly hardy as long as their habitat is preserved. Even attempts to eradicate insect pests are successful only on a local basis, and normal collecting techniques generally have no lasting effect on the population size.

To ensure permanency and to make collections of value beyond their aesthetic appearance, they must be properly prepared, labeled, and cared for. The references cited above will provide basic instructions for construction of equipment and preparation of specimens. Further suggestions for inexpensive materials and techniques, especially for beginners, are the following.

Finding specimens — A collector needs only diligence because insects occur virtually everywhere. The twenty microhabitats of insects listed in the introductory chapter indicate the kinds of places that should be searched. An excellent way to obtain night-flying insects is to take advantage of their tendency to be drawn to lights. An outdoor porch light, street lamp, or a Coleman lantern set on a bed sheet may attract hundreds of specimens. Most collectors use a 15-watt, 18-inch fluorescent ultra-violet light which is particularly attractive to nocturnal species.

Catching specimens — The most common collecting tools are aerial, aquatic, and sweep nets. Aerial nets should be made of fine-meshed material that is easy to see through, yet durable. A hoop 30 cm (12 inches) in diameter and a handle about 75 cm (30 inches) in length is a convenient size. The net bag should be long enough to fold completely across the face of the hoop in order to leave a small pocket in which the specimens are trapped. It should be affixed to a collar of stronger material such as muslin, through which the net rim is slipped. Aquatic nets also vary in size and shape; even a kitchen strainer will suffice for many situations.

Specimens should never be pinched or squashed. Instead, collectors use vials or bottles containing a volatile killing agent into which the specimens are captured directly or coaxed from the net. A standard vial contains potassium cyanide between layers of sawdust and held in place by plaster of paris or a tightly rolled strip of blotter paper. However, because this is such a dangerous poison, safer chemicals may be preferred, especially for children. Good killing bottles may be made by pressing a layer of cotton into the bottom, covering it with a disc of blotter paper, and soaking it with ethyl acetate, or a drop of lighter fluid can be applied directly to the insect's head. These are volatile fluids capable of killing insects, and are safer to handle. The disadvantage of such materials is the need to replenish frequently.

Corked or screw-cap Alka-Selzer or olive bottles make convenient killing jars, but the collector should have at least one larger bottle for big insects. Bottles should always be provided with some crumpled tissue paper in order to avoid damage to delicate specimens and to absorb moisture. Separate vials should be carried for Lepidoptera in order to minimize damage to their scale patterns and keep scales off other insects.

Digging and cutting tools, such as a trowel, knife with strong blade, or a lightweight crowbar, are useful for exhuming insects from the ground, under bark of dead trees, and similar situations. Other helpful implements include forceps and a hand lens. Relatively inexpensive, 10X or 12X lenses are available commercially. A shoulder bag is handy for carrying equipment; the hand lens and forceps can be leashed to the bag with string, and the forceps can be carried safely in a holster of rubber tubing.

To collectors of moths and other nocturnal insects, a portable ultraviolet light is indispensable. The GE "BL" Blacklight, a 15-watt, 18 inch, unfiltered fluorescent tube, is a convenient size. A 12-volt automobile battery can be used as a power source, either by means of a small converter available in campers' supply stores or by a transistorized fixture that uses direct current and is less drain on the battery. A bed sheet provided with a sleeve and a length of rope along one edge is needed. The sheet is simply hung like a clothesline between 2 trees or from the automobile to a fencepost, providing a landing surface for the insects which is convenient to collect from.

Transporting specimens. —Careful handling during temporary transport in the field will prevent damage to delicate specimens. Tight containers, such as snap-top plastic pill vials, lined with tissue are handy. They prevent drying, and the specimens may be pinned in the evening or the following day. A box with layers of tissue or cellucotton is needed if time does not permit pinning while specimens are fresh. Later they can be relaxed and pinned. A box such as a cigar box with soft cardboard or fiberboard pinning bottom to accommodate field-pinned specimens is handy if the collecting trip lasts more than a day. Glassine envelopes are often used to house butterflies, dragonflies, and a few other kinds of insects. Held in this manner, these specimens can be transported from the field for later pinning or can be identified and kept like a card file, which uses far less space than pinning and spreading them does.

Various-sized vials that can be filled with alcohol and stoppered to prevent evaporation are necessary for the preservation of soft-bodied specimens. Screw-cap vials are best for field use, while shell vials stoppered with cotton and inverted in alcohol in a bottle are preferable for storage because there is less evaporation. A wide-mouthed peanut butter jar with a rubber-lined cap serves nicely for this purpose. A 70 or 75% solution of ethyl alcohol or isopropyl (rubbing) alcohol is the best general purpose preservative. Soft larvae tend to turn black in alcohol, which can be prevented by killing them in boiling water and later transferring to alcohol.

Often it is desirable to transport living specimens for later study at home. Adults of many insects are best obtained by bringing larvae and their food (leaves, galls, mines, etc.) in and rearing them. Polyethylene bags make an excellent all-purpose field container. Often the rearing can be carried out in them. These should be larger and stronger than sandwich "baggies," which allow dessication. If the trip is a long one or traverses hot regions, the bags can be stored in a camp ice box, segregated from the ice by a trimmed-down cardboard box. They should be provided with paper toweling to absorb excess moisture. Clothespins make handy closures for the bags in the field.

Preparation and storage of specimens. —Insect pins are specially constructed to avoid rusting or corrosion and are longer than sewing straight pins. They are available in various diameters and lengths to accommodate varied body sizes of insects. Although very frail pins (size #0, 1) sometimes may be desired, size #2 and #3 are all-purpose sizes most often used. Mounting procedures vary according to the kind of insect. Most larger specimens are pinned through the midthorax, leaving about 10 mm below the pinhead as finger room for holding the specimens. Hemiptera are usually pinned through the scutellum, and Coleoptera always through the right elytron. Pinned and mounted specimens require no preservation treatment other than thorough drying. Kept dry and free of pests in a tight box, a collection of dried insects will last indefinitely.

Insects too small to be pinned directly through the body with insect pins are prepared by means of points or minutens. Points are slender triangles of stiff paper about 7 mm long and 2 mm wide at the base. They are pinned through the broad end with a #3 insect pin, and the specimen is glued to the tip of the point. A small crimp should be made in the tip with forceps to better accommodate the specimen. Clear fingernail polish makes a good point glue. Minutens are very small, steel pins without heads, which are used for pinning tiny insects onto blocks of cork or polyporus pith, which are in turn pinned on #3 insect pins.

Some specimens of moths, butterflies, dragonflies, grasshoppers, and occasionally other insects should have the wings of one or both sides spread so as to allow study of the whole surface. A spreading board is a long, slotted board, fitted so that several specimens can be pinned onto it at once. The wings are held in place for drying by pinning down paper strips over them. Spreading blocks for individual specimens of various sizes are easier to construct and more efficient to use. The block should have the wingspreading surfaces slightly tilted inward to the median slot and a hole drilled from the bottom to the slot, in which yucca stalk pith or other soft material is tightly fitted to receive the pin. Thread, held in place by razor-blade cuts at the margins of the block, is wrapped around 3 or 4 times on each side. The wings are first held loosely in place by the innermost thread wraparound. After positioning the wings with the tip of a pin, being careful not to puncture the wings, the remainder of the thread is wrapped snugly around the block. With Lepidoptera, care should be taken not to pull the thread too tightly, causing impressions on the wings. In normal indoor conditions, 4 or 5 days are sufficient for drying, after which the specimens can be removed from the blocks and the wings remain as positioned.

Specimens that have dried must be relaxed before they are mounted. Specimens placed in the relaxing chamber absorb moisture, become pliable, and can then be handled without breakage. A wide-mouth glass jar or plastic refrigerator box with snug lid can be used. Damp paper toweling is laid on the bottom and specimens are placed on a glass or plastic dish on the toweling for 1–2 days. Mold preventative such as phenol (carbolic acid), moth ball crystals, or chlorocresol must be provided. The relaxing chamber also provides a convenient storage place for freshly killed material overnight until time is available for pinning. In fact, Lepidoptera are more easily handled for spreading if they are stored in the relaxing chamber overnight than if they are pinned the same day as collected. If a few days or longer are likely to elapse, specimens can be stored in freezer dishes in a refrigerator freezer.

For display or storage, specimens should be housed in pest-proof containers. Most entomological institutions store their collections in glass-top drawers with unit trays of standard sizes, fitted with cork or polyethylene foam on the bottom. The trays can be shifted and arranged without the necessity of repinning specimens. For the beginner, cigar boxes make inexpensive storage boxes. Soft cardboard or fiberboard can be cut to fit and glued in as a pinning bottom. A heated pinhead pushed into a mothball provides a museum pest deterrent that can be stuck in a corner of the box.

To be of scientific value every insect specimen must have data labels permanently associated with it. Labels should be small, on good quality paper, and contain the following information. *Locality:* The place of collection should be given as exactly as practical and should be so designated that it can be found on a road map. *Date:* The day, month, and year. *Collector:* The collector should be indicated as a possible source of further information. *Ecological Source:* If known, the host, food plant, or other activity should be indicated (for example, "under loose bark pine stump," "mating on fungus 1030AM"). Labels can be neatly printed by hand, using a crowquill artist's pen and India ink. When needed in large numbers they can be inexpensively obtained by typing the information on white bond paper and having a photostat or offset negative produced by a copying and duplicating service at about ⅓ the original size. Copies of typewritten sheets with many labels can be made at a few cents each, from the negative.

Precautions against museum pests such as dermestid beetles and book lice, which eat dried specimens in insect collections, are a necessary part of the care of material not preserved in liquid. Naphthalene is satisfactory as a repellent, but to kill the pests, a fumigant such as Paradichlorobenzene (PDB) is necessary. Both of these are readily available as household "moth" preventatives for linen closets. Neglected collections may be reduced to dust within a few months.

GLOSSARY

alate   winged form.

annulation   ringlike segment or division.

anterior   towards front end.

apical   end, tip, outermost point.

arista   large, dorsal bristle on apical antennal segment of most Diptera, Cyclorrhapha.

basal   near the base or point of attachment.

campodeiform   shaped like a Dipluran, Campodeidae.

carrion   flesh of dead vertebrate animal.

caste   a form of adult in social insects.

caudal   towards the tail end.

cercus (pl. cerci)   one of a pair of appendages at the tail end of the abdomen.

chrysalis   pupa of a butterfly.

collophore   tubelike structure on the venter of the first abdominal segment in Collembola.

compound eye   eye composed of many elements (ommatidia).

cornicles   pair of tubular structures on the posterior part of aphid abdomen.

costa (costal)   leading edge of the wing (pertaining to that area of the wing).

coxa (pl. coxae)   basal segment of the leg.

crossvein   wing vein connecting adjacent longitudinal veins.

described species   formally classified and named.

distal   towards the free end of an appendage, end farthest from the body.

diurnal   active in the daytime.

dorsum (dorsal)   the back or upperside (pertaining to topside).

ectoparasite   parasitic insect which lives on the exterior of its host.

elytron (pl. elytra)   one of the thickened, often hardened forewings in Coleoptera.

endoparasite   parasite which lives inside its host.

epicranial suture   external groove in upper part of head, between face and neck.

exarate pupa   with appendages free, not glued to body.

exuvia (pl. exuviae)   cast skin of immature insect.

femur (pl. femora)   third leg segment, between the trochanter and tibia.

filiform   threadlike.

frass   excrement of insect.

furcula   forked, springing apparatus in Collembola at end of the abdomen.

gall   abnormal growth of plant tissues caused by insect or other agency such as mites.

globose   spherical.

grub   thick-bodied larva, blunt at both ends.

halter (pl. halteres)   small, knobbed structure representing the hindwing in Diptera.

hemelytron (pl. hemelytra)   forewing of Hemiptera, Heteroptera.

holometabolous   with complete metamorphosis, change in form (egg, larva, pupa, adult).

inquiline   species that lives in the home (nest, burrow, gall) of another species.

instar   stage between successive molts.

integument   outer covering of insect body.

keel   elevated ridge, often shaped like the keel of a boat.

labial palpus (pl. palpi)   segmented, antennalike structure arising on each side of labium.

labium   lower "lip" of the mouthparts.

labrum   upper "lip" of mouthparts.

lanceolate   spear-shaped, tapering to the tip.

larva (pl. larvae)   immature, feeding stage between egg and pupa of holometabolous insects.

larviform   shaped like a larva.

lateral   referring to the side of the body or individual structure.

maggot   legless, headless larva of Diptera, often carrot-shaped, blunt at the posterior end.

mandible   one of lateral, opposed structures of mouthparts, usually adapted for biting.

mandibulate   with mandibles.

maxilla (pl. maxillae)   one of paired lateral mouthpart structures just behind the mandibles.

maxillary palpus (pl. palpi)   segmented, antennalike structure arising from maxilla.

mesothorax   middle segment of the thorax, bearing forewings in winged insects.

metathorax   hind or third segment of thorax, bearing hindwings in winged insects.

miner   larva living within plant leaves or in stems just under the epidermis.

morphology   study of form or structures.

naiad   aquatic, gill-breathing immature stage of Odonata, Plecoptera, and Ephemeroptera.

notum   dorsal surface of a body segment.

nymph   immature, feeding stage between the egg and adult in nonholometabolous insects.

obtect pupa   with the appendages affixed to the body surface.

ocellus (pl. ocelli)   simple eye, usually located behind the compound eye; in a group of three in some insects.

ommatidium   single unit of a compound eye.

ovipositor   egg-laying apparatus, modified as a sting in many Hymenoptera.

palpus (pl. palpi)   segmented, antennalike process of labium or maxilla.

parasite   species that feeds in or on another organism (its host) during all or part of its life but does not normally kill the host; in insects often refers also to parasitoid species.

parasitoid   species that lives in another insect and kills it.

pecten   comblike or rakelike structure.

petiole (pedicel)   "stem" or narrow basal part of abdomen in Hymenoptera.

pheromone   chemical given off by an individual, causing reaction in other individuals of the same species.

phytoplankton   plants drifting in marine or fresh water.

plumose   featherlike.

postclypeus   upper portion of the clypeus or "face."

posterior   referring to tail end.

prehensile   adapted for grasping or holding.

proboscis   tube-shaped process of the head; modified mouthparts of some insects.

process   outgrowth or appendage of body wall or structure.

proleg   abdominal "false legs" of larvae; shortened name for prothoracic legs of adults.

pronotum   dorsal sclerite of prothorax, often enlarged, extending over mesothorax.

propodeum   first abdominal segment in Hymenoptera, which is united with the thorax.

prothorax   anterior or first thoracic segment.

pubescence (pubescent)   covered with short, fine hairs.

pupa (pl. pupae)   immature stage after the larva in which transformation to the adult stage occurs in holometabolous insect.

puparium   case formed by hardening of the larval skin, in which pupation takes place in most Diptera, Cyclorrhapha.

raptorial   modified for grasping prey.

reticulate   like a network.

rostrum   beak.

rugose   with many wrinkles or small ridges.

sclerite   hardened body wall plate.

sclerotized   hardened.

scutellum   sclerite of the thoracic dorsum, usually the mesothoracic.

spiracle   breathing pore, external opening of a trachea.

stria (pl. striae)   thin groove or depressed line.

stridulate (stridulatory)   to make a noise by rubbing two surfaces together (referring to structure used for this purpose).

stylet   needlelike structure, especially piercing structures of sucking mouthparts.

stylus (pl. styli)   fingerlike process of abdomen.

tarsus (pl. tarsi)   last or distal leg segment, consisting of one to five subdivisions or "segments."

tegmen (pl. tegmina)   thickened, leatherlike forewing of an Orthopteran.

tibia (pl. tibiae)   fourth leg segment, beyond the femur and just preceding the tarsus.

trochanter   second segment of a leg, between the coxa and femur.

tubercle (tuberculate)   knoblike or rounded process (with tubercles).

venter (ventral)   bottom or underside.

vestiture   body covering of hairs or scales.

wing covers   elytra.

wing pad   precursor of adult wing in nymph.

xeric   characterized by scanty moisture, referring or adapted to arid conditions.

ABBREVIATIONS

BL   = body length

C   = centigrade

ft   = feet

FW   = forewing

HW   = hindwing

in   = inches

km   = kilometers (1 km = about 0.6 miles)

m   = meters (1 m = about 3.25 feet)

mi   = miles

mm   = millimeters (25 mm = about 1 inch)

p. (pp.) = page (pages)

pl.   = plural

sp.   = species (singular)

spp.   = species (plural)

WL   = wing length

INDEX